Heat and Mass Transfer

Series Editors

Dieter Mewes, Universität Hannover, Hannover, Germany

Franz Mayinger, München, Germany

This book series publishes monographs and professional books in all fields of heat and mass transfer, presenting the interrelationships between scientific foundations, experimental techniques, model-based analysis of results and their transfer to technological applications. The authors are all leading experts in their fields.

Heat and Mass Transfer addresses professionals and researchers, students and teachers alike. It aims to provide both basic knowledge and practical solutions, while also fostering discussion and drawing attention to the synergies that are essential to start new research projects.

More information about this series at http://www.springer.com/series/4247

Christo Boyadjiev

Editor

Modeling and Simulation in Chemical Engineering

Project Reports on Process Simulation

 Springer

Editor
Christo Boyadjiev
Institute of Chemical Engineering
Bulgarian Academy of Sciences
Sofia, Bulgaria

ISSN 1860-4846   ISSN 1860-4854   (electronic)
Heat and Mass Transfer
ISBN 978-3-030-87662-3   ISBN 978-3-030-87660-9   (eBook)
https://doi.org/10.1007/978-3-030-87660-9

This Springer imprint is published by the registered company Springer Nature Switzerland AG
The registered company address is: Gewerbestrasse 11, 6330 Cham, Switzerland

This book is dedicated to the Institute
of Chemical Engineering, Bulgarian Academy
of Sciences.

Preface

A theoretical analysis of the methods for chemical engineering processes modeling is presented. The methods for modeling specific processes may be different, but in all cases, they must bring the mathematical description closer to the real process by using appropriate experimental data. These methods are presented in the cases of co-current absorption column without packings, counter-current absorption column with random packings, and modeling of processes with unknown mechanism.

The successful ethanol fermentation from lignocellulosic substrates is impeded by the fact that cellulose and hemicellulose must be digested by different strains; hence full substrate utilization is not possible by one microbial strain in one process. That is why these two polysaccharides should be treated separately by different microbial strains. The simplest way to accomplish these processes is to use a bioreactor with separated compartments where cellulose and hemicellulose are fermented by different strains: *Saccharomyces cerevisiae* for hexoses from cellulose and *Pichia stipitis* for pentoses from hemicellulose. In the present work such a two-step process is modeled for a continuous operation. As substrates glucose (a hexose) and xylose (a pentose) resulting of acid pretreatment of the ligno-cellulose are used. The addition of the produced ethanol in the first compartment to the second one as starting substrate and its inhibition impact are taken into account. The dilution rate, the initial substrate concentration, and the inhibition effects are considered too. Mathematical models describing these fermentation processes are composed and used for kinetic parameter estimation from own experimental data. Both product and substrate inhibitions are taken into account. It is demonstrated that the simultaneous glucose and xylose fermentation to ethanol by the strain *Pichia stipitis* is successfully described by the composed model. The conversion from monosaccharides to ethanol takes place with a yield of 70%.

A new approach for the chemical processes modeling in column apparatuses is presented in industrial column apparatuses. An exact approach for solutions of the equations in the convective-type models is used. The use of experimental data, for the average concentration at the column end, for a concrete process and column, permits to be obtained the model parameters, related with the radial non-uniformity of the velocity. These parameter values permit to be used the average-concentration model for modeling of chemical processes with different reaction rates.

The designing mathematical model of an integrated bioethanol supply chain (IBSC) that will account for economic and environmental aspects of sustainability is presented. A mixed integer linear programming model is proposed to design an optimal IBSC. Bioethanol production from renewable biomass has experienced increased interest in order to reduce Bulgarian dependence on imported oil and reduce carbon emissions. Concerns regarding cost efficiency and environmental problems result in significant challenges that hinder the increased bioethanol production from renewable biomass. The model considers key supply chain activities including biomass harvesting/processing and transportation. The model uses the delivered feedstock cost, energy consumption, and GHG emissions as system performance criteria. The utility of the supply chain simulation model is demonstrated by considering a biomass supply chain for a biofuel facility in Bulgarian scale. The results show that the model is a useful tool for supply chain management, including selection of the optimal bioethanol facility location, logistics design, inventory management, and information exchange.

The integration of energy and mass processes is one of the most powerful tools for creating sustainable and energy-efficient production systems. Process integration covers a wide range of system-oriented methods and approaches that are used in the design and reconstruction of industrial processes to obtain optimal use of resources. Traditionally, methods have focused on energy efficiency, but more recently they have also covered other areas, such as the integration of mass processes for the efficient use of water and other resources. In production systems with batch processes, the task of energy integration is significantly more complex due to the presence of predominantly low-potential heat, which until recently was considered not to be recoverable. In addition, the periodic and discrete nature of heat sources and recipients imposes additional constraints that require process coordination. Two main approaches for thermal integration of periodic processes are defined, direct and indirect:

1. Direct heat integration determines the existence of heat exchange between technological flows that occur simultaneously over time. This approach to heat recovery requires adherence to a strict production schedule to ensure energy efficiency and product quality. In the chapter different variants for direct heat integration are considered, with recirculation of the main fluids, or with the use of intermediate heating and cooling agents. The corresponding mathematical models are derived.

2. Indirect heat integration determines the existence of heat exchange between flows that do not occur simultaneously in the system. This approach uses intermediate fluids and a heat storage system (heat tanks) so that the heat can be stored, transferred, and utilized in a future period of time. It allows the heat exchange process to be less limited and less sensitive than schedules and provides some

operational flexibility. The possibilities for heat integration with the use of one common cold/hot heat tank and two separate ones but with a common heating/cooling agent are considered. The respective mathematical models are also presented.

Traditional approaches to modeling real chemical engineering processes are based on fundamental chemical and physical laws, which include nonlinear algebraic and differential equations. From a computational point of view, these equations have some difficulties with regard to the numerical methods used for their approximation, as well as with the achievement of the desired accuracy of the calculations. In recent years, there has been a growing interest in the application of the Artificial Neural Networks (ANNs) method to solve a number of problems in the field of chemical engineering related to fault detection, signal processing, modeling, and control of chemical and biochemical processes in which traditional modeling methods have difficulty and it is even impossible to develop physical models with acceptable errors. Their main advantage is that they work only with data on the input and output values of the process parameters. One model can be used to generate multiple outputs. Once the neural network model is adequately trained and validated, it is able to make predictions for new data about the input values of process parameters that were not used in the development of this model. This chapter presents the main characteristics of ANNs, the choice of architecture, the process of training and validation of ANN models, as well as several types of ANNs, such as feed-forward nets, recursive nets, and radial basis function nets and combined models with examples of applications in chemical engineering.

Solar thermal energy is of intermittent and dynamic character, and efforts to use this energy during non-sunshine periods are one of the current interests of researchers. The research for the development of thermal energy accumulators has prompted researchers towards the creation of compact solutions using latent heat storage. Phase change materials as thermal energy accumulators are attractive because of their high storage density and characteristics to release thermal energy at constant temperature corresponding to the phase transition temperature. The present chapter shows the recent state of the art on small-scale solar thermal dryers integrated with phase change material as energy accumulators. This is an intensive field of investigation for more than 30 years with importance for agriculture and food industry especially in hot climate. A variety of commercial small-scale solar dryers are offered as a low-cost, zero-energy solution for small farmers. And yet, there are no commercial systems using latent thermal storage because at the present level of development this unit will increase too much the price of the system, which contradicts its purpose. The solution needs very simple design, accessible materials, and optimal conditions for operation. The aim of the present chapter is to makes an overview of the methods for theoretical evaluation and prediction, which are used to design and assess these devices and to find the most appropriate of them for this new solution. On the bases of modeling and simulation, it compares the most cost- and energy-effective solar dryer systems with thermal storage among the great number of designs, devices, and materials. The resulting conclusions from the collected and compared information will serve as prerequisites for a novel solution of a cost-effective thermal energy storage for a small-scale solar dryer, which will lead to improved efficiency of the drying process, due to controlled temperature and longer operation time. This information might serve also in the development of the wider field of thermal energy storage, which is an important part of the technologies of renewable and waste energy conversion.

Sofia, Bulgaria   Christo Boyadjiev

Motto

*The **mathematical models** of the industrial processes*
*are **relations** between the **physical models***
*and **mathematical descriptions** by*
experimental data.

Contents

Chapter 1
Introduction in the Chemical Engineering Processes Modeling

Christo Boyadjiev

Abstract In the paper is presented a theoretical analysis of the methods for chemical engineering processes modeling. The methods for modeling specific processes may be different, but in all cases they must bring the mathematical description closer to the real process by using appropriate experimental data. These methods are presented in the cases of co-current absorption column without packings, counter-current absorption column with random packings and modeling of processes with unknown mechanism.

Key words Modeling · Physical model · Mathematical description · Experimental data · Process kinetics · Thermodynamic approximation · Hydrodynamic approximation · Boltzmann's approximation · Absorption models · Guchmann's theorem

1 Prelude

The main problems in the chemical industry (biotechnology, heat energy) are the optimal design of new devices and the optimal control of active processes, i.e., minimization of the investment and operating costs. These problems are solved by chemical engineering with modeling methods [1].

The creation of the mathematical model begins with the formulation of the physical model of the complex process, i.e., the definition of the simple processes that make it up and the interactions between them. The second step is to define simple processes that have mathematical descriptions (equivalent mathematical operators). The other simple processes are introduced into the mathematical model through quantitative information obtained from experimental data, which brings the mathematical model as close as possible to the real process. The experiment brings mathematics closer to physics (reality).

C. Boyadjiev (✉)
Institute of Chemical Engineering, Bulgarian Academy of Sciences, Sofia, Bulgaria

The optimal design and control in the chemical industry is uniquely related to processes rates, so all mathematical descriptions of processes are linked to algorithms to determine these rates, i.e., processes kinetics.

2   Industrial Processes Kinetics

The industrial systems consist of separate phases (gas, liquid, solid) in the industrial apparatuses volumes. They are in thermodynamic equilibrium when the velocities, temperatures, and concentrations of substances in the individual parts or points of the phases are equal.

The processes in the chemical industry (biotechnology, heat energy) are a result of the deviation of the systems from their thermodynamic equilibrium [2]. One system is not in a thermodynamic equilibrium when the velocities, concentrations of the components (substances), and the temperatures at the individual points in the phase volumes are different. These differences are the result of reactions, i.e., of processes that create or consume substance and (or) heat. As a result, the industrial processes kinetics is equivalent to the reactions kinetics [3].

The presented analysis shows that processes in the chemical industry are result of reactions that occur in the phase volume (homogeneous) or on the boundary between two phases (heterogeneous). Homogeneous reactions are generally chemical, while heterogeneous reactions are chemical, catalytic, physical and chemical adsorption, interphase mass transfer in gas-liquid and liquid-liquid systems (on the interphase surface the substance disappears from one phase and occurs in the other phase). The rates of these processes are determined by the reaction kinetics [3], which lies at the basis of modeling in chemical engineering, and solving the basic problems in the chemical industry (biotechnology, heat energy).

3   Modeling

The basics of modeling in chemical engineering, as part of human knowledge and science, are related to the combination of intuition and logic that has different forms in individual sciences [4]. In the mathematics, the intuition is the axiom (unconditional statements that cannot be proven), while the logic is the theorem (the logical consequences of the axiom), but logic prevails over intuition. In the natural sciences (physics, chemistry, biology), the "axioms" (principles, postulates, laws) are not always unconditional, but logic prevails over intuition too.

The processes in chemical engineering take place in the industrial apparatuses, where gas, liquid, and solid phases move together or alone. They are described by variables, which are extensive or intensive. In the case of merging of two identical systems, the extensive variables are doubled, but the intensive variables are retained.

In the chemical industry (biotechnology, heat energy), processes take place in moving phases (gas, liquid, solid). Reactions (reaction processes) lead to different concentrations (and temperatures) in the phase volumes and the phase boundaries. As a result, hydrodynamic processes, diffusion mass transfer, and heat conduction are joined to the reaction processes. Under these conditions there are various forms of mass transfer (heat transfer) that are convective (as a result of phase movements) and diffusion (as a result of concentration (temperature) gradients in the phases).

Convective mass transfer (heat transfer) can be laminar or turbulent (as a result of large-scale turbulent pulsations). Diffusion mass transfer (heat transfer) can be molecular or turbulent (as a result of small-scale turbulent pulsations).

Mathematical models of industrial apparatuses aim at determining the concentration of substances (flow temperatures) in the phases. They have different degrees of approximation—thermodynamic, hydrodynamic, and Boltzmann's approximations.

4 Thermodynamic Approximation

The processes in chemical engineering are the result of a deviation from the thermodynamic equilibrium between two-phase volumes or the volume and phase boundaries of one phase and represent the pursuit of systems to achieve thermodynamic equilibrium [2]. They are irreversible processes and their kinetics use mathematical structures derived from Onsager's principle of linearity. According to him, the average values of the derivatives at the time of the extensive variables depend linearly on the mean deviations of the conjugated intensive variables from their equilibrium states. The principle is valid close to equilibrium, and the Onsager's linearity coefficients are kinetic constants. When the process is done away from equilibrium (high intensity processes) kinetic constants become kinetic complexes, depending on the corresponding intensive variables. The thermodynamic approximation models cover the entire volume of the phase or part of it.

5 Hydrodynamic Approximations

The hydrodynamic level uses the approximations of the mechanics of continua, where the mathematical point is equivalent to an elementary physical volume, which is sufficiently small with respect to the apparatus volume, but at the same time sufficiently large with respect to the intermolecular volumes in the medium. In this level the molecules are not visible, as is done in the next level of detail of Boltzmann.

The models of the hydrodynamic approximations are possible to be created on the basis of the mass (heat) transfer theory, whose models are created by the models of the hydrodynamics, diffusion, thermal diffusion and reaction kinetics, using the logical structures of three main "axioms," related with the impulse, mass, and heat transfer:

1. The postulate of Stokes for the linear relationship between the stress and deformation rate, which is the basis of the Newtonian fluid dynamics models.
2. The first law of Fick for the linear relationship between the mass flow and the concentration gradient, which is the basis of the linear theory of the mass transfer.
3. The first law of Fourier for the linear relationship between the heat flux and the temperature gradient, which is the basis of the linear theories of the heat transfer.

These are the laws of the impulse, mass, and energy transfer.

6   Boltzmann's Approximation

In Boltzmann's kinetic theory of the ideal gas, the hydrodynamic "axioms" are three "theorems" that derive from the axiom of the "elastic shock" (in a shock between two molecules the direction and the velocity of the movement change, but the sum of their kinetic energies is retained, i.e., there is no loss of kinetic energy) and the rate coefficients are theoretically determined by the average velocity and the average free run of the molecules.

7   Mechanism of Influence of Reaction Kinetics

The mathematical model of an engineering chemical process is a mass (heat) balance in the phase volumes, where the mathematical operators are mathematical descriptions of the composite processes, and the relationship between them (differential equations) corresponds to the mechanism of the complex process. The boundary conditions of the differential equations are formulated at the interphase boundaries. For this purpose, the knowledge of the mathematical descriptions of the velocity distribution in the phases and the interphase boundaries is necessary.

Industrial processes are a set of physical and chemical reactions, hydrodynamic, diffusion and thermal processes that take place in the industrial apparatus volume. The problems in compiling the models of the kinetics of industrial apparatuses arise from the need for information about the interaction between the individual processes in the complex process (its mechanism) and a mathematical description of the geometry of the industrial apparatus volume.

For the most part, industrial cases do not have the above information, which requires simplification of the models and introduction of some effects through experimentally determined parameters. As examples will be considered a co-current absorption column without packings and a counter-current absorption column with random packings.

8   Co-Current Absorption Column without Packings

In the absorption columns without packings, the velocity distributions in the gas and liquid phases and the interfacial limits are unknown, i.e., the differential equations (mass balances in the phases) and their boundary conditions at the interphase boundaries (velocity of the interphase mass transfer) cannot be formulated. These problems are overcome by creating of convection-diffusion and average-concentration models [5, 6]. In the convection-diffusion model, the velocity of the interphase mass transfer is replaced by volume physical reaction and experimentally determinable parameter. In this model, the velocities are unknown, so it can only be used for qualitative analysis. From it the average-concentration models are obtained, by model averaging along the cross section of the column. The obtained average-concentration model involves average velocities and concentrations, and the velocity distributions in the phase volumes are introduced with experimentally determined parameters.

8.1   Convection-Diffusion Model

In the stationary case, the convection-diffusion model [3, 4] of the co-current chemical absorption process, with a pseudo-first-order chemical reaction in the liquid phase, in cylindrical coordinate system (r, z) [m], has the form:

$$u_j \frac{\partial c_j}{\partial z} + v_j \frac{\partial c_j}{\partial r} = D_j \left(\frac{\partial^2 c_j}{\partial z^2} + \frac{1}{r} \frac{\partial c_j}{\partial r} + \frac{\partial^2 c_j}{\partial r^2} \right) + (-1)^{(2-j)} k(c_1 - \chi c_2) - (j-1)k_0 c_2;$$

$$r = 0, \quad \frac{\partial c_j}{\partial r} \equiv 0; \quad r = r_0, \quad c_j = (j-1)c_1^0 \chi^{-1}; \quad j = 1, 2;$$

$$z = 0, \quad c_1 \equiv c_1^0, \quad c_2 \equiv 0, \quad u_1^0 c_1^0 \equiv u_1 c_1^0 - D_1 \left(\frac{\partial c_1}{\partial z} \right)_{z=0}, \quad \left(\frac{\partial c_2}{\partial z} \right)_{z=0} = 0.$$

$$(1.1)$$

$$\frac{\partial u_j}{\partial z} + \frac{\partial v_j}{\partial r} + \frac{v_j}{r} = 0;$$

$$r = r_0, \quad v_j(r_0, z) = 0; \quad z = 0, \quad u_j = u_j(r, 0); \quad j = 1, 2.$$

$$(1.2)$$

In (1.1 and 1.2) $u_j = u_j(r, z)$, $v_j = v_j(r, z)$ and $c_j = c_j(r, z)$ are the axial and radial velocity components and transferred substance concentrations in the gas ($j = 1$) and liquid ($j = 2$) phases, D_j are the diffusivities in the phases, u_j^0 and c_j^0 are the inlet velocities and the concentrations in the phases, k is the interphase mass transfer rate coefficient, χ—the Henry's number, k_0 the chemical reaction rate constant. The concentrations of the transferred substance in the phases are presented as kg-mol of the transferred substance in 1 m^3 of the phase volume. The inlet velocities u_j^0

$(j = 1, 2)$ of the gas and liquid phases are equal to the average velocities $\bar{u}_j\,(j = 1, 2)$ of the phases in the column.

On the column wall the velocity components are zero $(r = r_0,\ u_j = v_j \equiv 0,\ j = 1, 2)$, i.e., there is no convective mass transfer. At the surface of the column, the motionless gas phase substance is absorbed into the motionless liquid phase. As a result, the concentration of the absorbent substance in the gas on the wall decreases to zero and its concentration in the liquid increases maximally (until thermodynamic equilibrium is reached), i.e., $r = r_0,\ c_1 = 0,\ c_2 = c_1^0\chi^{-1}$.

In the physical absorption, the interphase mass transfer between gas and liquid phases is a surface physical reaction. In (1.1) this reaction is presented as a volume reaction and its rate $Q = (-1)^{(2 - j)}k(c_1 - \chi c_2),\ j = 1, 2$ participates in the mass balances in the gas and liquid phases.

8.2 Average-Concentration Model

The averaging of the convection-diffusion model [5, 6] along the cross section of the column leads to the average-concentration model:

$$\alpha_j(z)\bar{u}_j\frac{d\bar{c}_j}{dz} + \left[\beta_j(z) + \varepsilon\gamma_j(z)\right]\bar{u}_j\bar{c}_j =$$
$$= D_j\frac{d^2\bar{c}_j}{dz^2} + (-1)^{(2-j)}k(\bar{c}_1 - \chi\bar{c}_2) - (j-1)k_0\bar{c}_2; \qquad (1.3)$$
$$z = 0, \quad \bar{c}_j(0) \equiv (2-j)c_j^0, \quad \frac{d\bar{c}_j}{dz} \equiv 0; \quad j = 1, 2,$$

where

$$\alpha_j(z) = \frac{2}{r_0^2}\int_0^{r_0} r\frac{u_j c_j}{\bar{u}_j\bar{c}_j}\,dr, \quad \beta_j(z) = \frac{2}{r_0^2}\int_0^{r_0} r\frac{u_j}{\bar{u}_j\bar{c}_j}\frac{\partial c_j}{\partial z}\,dr,$$
$$\gamma_j(z) = \frac{2}{r_0^2}\int_0^{r_0} r\frac{v_j}{\varepsilon\bar{u}_j\bar{c}_j}\frac{\partial c_j}{\partial r}\,dr, \quad \varepsilon = \frac{r_0}{l}. \qquad (1.4)$$

The functions $\alpha_j(z),\ \beta_j(z),\ \gamma_j(z),\ j = 1, 2$ are possible to be presented as next approximations:

$$\alpha_j(z) = 1 + a_{j1}z + a_{j2}z^2, \quad \beta_j(z) = 1 + b_{j1}z + b_{j2}z^2,$$
$$\gamma_j(z) = 1 + g_{j1}z + g_{j2}z^2, \quad j = 1, 2, \qquad (1.5)$$

where the values of a_{j1}, a_{j2}, b_{j1}, b_{j2}, g_{j1}, g_{j2}, $j = 1, 2$ is possible to be obtained, using experimental data for the average concentration at the column end:

$$\bar{c}_{jm}(l), \quad j = 1, 2, \quad m = 1, \ldots, 10. \tag{1.6}$$

The introducing of (1.5) into (1.3) leads to

$$\left(1 + a_{j1}z + a_{j2}z^2\right)\bar{u}_j \frac{d\bar{c}_j}{dz} + \left[\left(1 + b_{j1}z + b_{j2}z^2\right) + \varepsilon\left(1 + g_{j1}z + g_{j2}z^2\right)\right]\bar{u}_j\bar{c}_j =$$

$$= D_j \frac{d^2\bar{c}_j}{dz^2} + (-1)^{(2-j)}k(\bar{c}_1 - \chi\bar{c}_2) - (j-1)k_0\bar{c}_2;$$

$$z = 0, \quad \bar{c}_j(0) \equiv (2-j)c_j^0, \quad \frac{d\bar{c}_j}{dz} \equiv 0; \quad j = 1, 2,$$

$$\tag{1.7}$$

where the unknown parameters values $P(k, a_{j1}, a_{j2}, b_{j1}, b_{j2}, g_{j1}, g_{j2}, j = 1, 2)$ must be obtained, using experimental data, by the minimization of the least-squares function Q with respect to P:

$$Q(P) = \sum_{m=1}^{10} [\bar{c}_1(l, P) - \bar{c}_{1m}(l)]^2 + \sum_{m=1}^{10} [\bar{c}_2(l, P) - \bar{c}_{2m}(l)]^2, \tag{1.8}$$

where $\bar{c}_1(l, P)$, $\bar{c}_2(l, P)$ are solutions of the average-concentration model (1.7).

The presented approach is used for modeling of chemical, absorption, adsorption, and catalytic processes in column apparatuses without packings [5, 6].

9   Counter-Current Absorption Column with Random Packings

Counter-current absorption column with random packings are characterized by the presence of a layer of liquid that flows along the wall of the column and practically does not participate in the absorption process and reduces the working volume of the column. The created hydrodynamic situation does not allow the approach used in the modeling of columns without packings.

9.1   Fluid Flow along the Column Wall

The liquid flow on the surface of the random packings and when it reaches the column wall most of it flows on this surface and cannot return to the column volume due to the small contact surface between the wall surface and the random packings. The thickness of the flowing layer of liquid increases and conditions are created for the return of liquid from the layer to the packings and further the two effects are equalized. In this way, the layer of liquid reaches a constant maximum thickness, with which it moves to the end of the column. The amount of liquid entering the flowing layer leads to a reduction in the amount of liquid in the volume of the column, i.e., to the radial non-uniformity of the axial component of the liquid velocity in the column and to the reduction of the mass transfer rate in the liquid phase. In addition, this layer is not involved in the absorption process.

The effect of liquid flowing on the column wall is the result only of the geometric shape of the random packings and thus determines the rate of absorption of slightly soluble gases, which reaches a maximum value at maximum packings surface per unit volume of the column and minimum thickness of the flowing layer of liquid.

9.2   Problems with Random Packings in the Columns

In the case of modeling the hydrodynamics in the gas and liquid phases in columns with random packings, the following problems arise:

1. The flow rate of the liquid flowing on the surface of the random packings [$m^3 \cdot s^{-1}$] and the retention of the liquid on this surface are unknown.
2. The flow rate of the liquid flowing on the column wall [$m^3 \cdot s^{-1}$] and the retention of the liquid on this surface are unknown.
3. The hydrodynamic resistance of the random packings on the gas phase is unknown.

Theoretical analysis [7] shows that this problem can be overcome in the presence of experimental data on the flow rate of the liquid flowing along the column wall, at different packings heights and on the packing pressure drop, during the movement of the gas phase. This requires the creation of a hydrodynamic model in which the liquid and the gas move in separate channels and interact on the surface of the flowing layer. The introduction of experimentally determined quantities into the model brings it as close as possible to the real process.

9.2.1   Experimental Data

The wall flow thickness in columns with random packing changes smoothly from 0 to a constant maximal value δ_{max} and can be expressed by an approximation function $\delta(z)$:

$$\delta(z) = \frac{z}{a + bz}, \quad \delta(0) = 0, \quad \delta(\infty) = \frac{1}{b} = \delta_{max}, \tag{1.9}$$

where z is the axial coordinate. The parameters (a, b) are possible to be determined from experimental data for the flow rate of the wall flow Q_{WF} at various packing heights l in the column - $Q_{WF}(z)$, $z = l_i$, $i = 1, \ldots, n$. The available experimental data can be described by the following approximation:

$$Q_{WF}(z) = \frac{z}{k_1 + k_2 z}. \tag{1.10}$$

From (1.10) is possible to determine the flowrate $Q(z)$ of the wall flow per unit periphery of the column $(2\pi r_0)$.

$$Q(z) = \frac{Q_{WF}(z)}{2\pi r_0} = \frac{z}{m_1 + m_2 z}, \quad m_1 = 2\pi r_0 k_1, \quad m_2 = 2\pi r_0 k_2, \tag{1.11}$$

where r_0 is the column radius.

9.2.2 Phase Volume Parts in the Column Volume

The volumes of the solid, gas, and liquid phases per unit volume of the column can be represented as:

$$\varepsilon_j, \quad j = 0, 1, 2, \quad \varepsilon_0 + \varepsilon_1 + \varepsilon_2 = 1, \tag{1.12}$$

where the indices $j = 0, 1, 2$ corresponds to solid, gas, and liquid phases.

As a result of the liquid flow on the column wall, the liquid phase is divided into two parts:

$$\varepsilon_2 = \varepsilon_{21}(z) + \varepsilon_{22}(z), \tag{1.13}$$

where $\varepsilon_{22}(z)$ is the fraction of the liquid flowing along the wall.

The gas and liquid flow rates Q_G, Q_L, Q_{WF} [m^3·s^{-1}] permit to be obtained the gas-liquid and liquid-liquid ratios in the column volume:

$$\bar{\varepsilon}_1 = \frac{Q_G}{Q_G + Q_L}, \quad \bar{\varepsilon}_2 = \frac{Q_L}{Q_G + Q_L}, \quad \bar{\varepsilon}_{21} = \frac{Q_L - Q_{WF}}{Q_L}, \quad \bar{\varepsilon}_{22} = \frac{Q_{WF}}{Q_L} \tag{1.14}$$

and gas and liquid hold-up in the packing:

$$\varepsilon_1 = \bar{\varepsilon}_1(1 - \varepsilon_0), \quad \varepsilon_2 = \bar{\varepsilon}_2(1 - \varepsilon_0), \quad \varepsilon_{21} = \bar{\varepsilon}_{21}\varepsilon_2, \quad \varepsilon_{22} = \bar{\varepsilon}_{22}\varepsilon_2. \tag{1.15}$$

The parameters ε_1, ε_2 denote the volume fraction, as well as the cross-section fraction, of the gas and liquid in the packing and are used to determine their inlet average velocities:

$$u_z^0 = \frac{Q_G}{\varepsilon_1 \pi r_0^2}, \quad v_z^0 = \frac{Q_L}{\varepsilon_2 \pi r_0^2}, \tag{1.16}$$

where u_z^0, v_z^0 are the average velocities in the void cross section of the gas and liquid phase at the inlet of the packing bed, Q_G, Q_L—gas and liquid inlet flow rates.

9.2.3 Pressure Drop of Random Packings Columns

The hydraulic resistance of the fillings H, i.e., the pressure drop through a layer of random packings with a thickness of 1 meter, at a given gas velocity:

$$H = \frac{p^0 - p(0, z_0)}{z_0}, \tag{1.17}$$

is determined from experimental data on the pressure difference at both ends $p^0 - p$ $(0, z_0)$ of the random packings height z_0.

The obtained results permit to be used a physical model, where the gas and the liquid move in parts of the column volume and through parts of the column cross section (ε_1, ε_{21}, ε_{22}) and contact on a cylindrical surface with variable radius $R_0(z) = r_0 - \delta(z)$.

9.2.4 Hydrodynamics of the Liquid Phase in the Column Volume

The flows in the column are axially symmetrical and the model of the hydrodynamics of the liquid phase in the volume of the column will be presented in a cylindrical coordinate system (r, z), where r and z are the radial and axial coordinates. In the packings columns, the pressure is constant during the movement of liquid under the action of weight. In this case the axial and radial components of the velocity v_z and v_r satisfy the Navier and Stokes equations:

$$v_z \frac{\partial v_z}{\partial z} + v_r \frac{\partial v_z}{\partial r} = \nu \left(\frac{\partial^2 v_z}{\partial z^2} + \frac{\partial^2 v_z}{\partial r^2} + \frac{1}{r} \frac{\partial v_z}{\partial r} \right) + g,$$
$$v_z \frac{\partial v_r}{\partial z} + v_r \frac{\partial v_r}{\partial r} = \nu \left(\frac{\partial^2 v_r}{\partial z^2} + \frac{\partial^2 v_r}{\partial r^2} + \frac{1}{r} \frac{\partial v_r}{\partial r} - \frac{v_r}{r^2} \right). \tag{1.18}$$

The inlet boundary conditions are:

$$z = 0, \quad v_z = v_z^0, \quad \frac{\partial v_z}{\partial z} = 0, \quad v_r = 0, \quad \frac{\partial v_r}{\partial z} = 0. \qquad (1.19)$$

The boundary conditions along the axis of the column are:

$$r = 0, \quad \frac{\partial v_z}{\partial r} = 0, \quad \frac{\partial v_r}{\partial r} = 0. \qquad (1.20)$$

The effect of the flow of liquid on the wall of the column must be taken into account when $r = R_0(z) = r_0 - \delta(z)$, where the amount of liquid which enters the wall of the column through the surface $2\pi R_0 dz$ by the radial velocity component v_r, i.e., $2\pi R_0 v_r dz$ must be equal to the volume of the liquid layer obtained on the wall of the column $2\pi r_0 v_0 d\delta$, where $v_0 = v_0(z)$ is the surface velocity of the liquid layer:

$$2\pi R_0 v_r (R_0, z) dz = 2\pi r_0 v_0(z) d\delta. \qquad (1.21)$$

As a result

$$r = R_0 = r_0 - \delta(z), \quad v_r(R_0, z) = \frac{r_0 v_0(z)}{r_0 - \delta(z)} \frac{d\delta}{dz}, \quad \frac{d\delta}{dz} = \frac{a}{(a + bz)^2}. \qquad (1.22)$$

The axial component of the liquid velocity v_z at the boundary $r = r_0 - \delta(z)$ must be equal to the surface velocity of the film flowing down the column wall $v_0(z)$:

$$r = r_0 - \delta(z) = r_0 - \frac{z}{a + bz}, \quad v_z = v_0(z), \qquad (1.23)$$

where the surface velocity changes smoothly from 0 to a constant maximal value and can be expressed by an approximation function:

$$v_0(z) = \frac{z}{\alpha + \beta z}. \qquad (1.24)$$

The parameters (a, b, α, β) in the approximation functions (1.9 and 1.24) must be obtained, using experimental data.

Finally, the hydrodynamic model of the liquid phase in column volume can be written as:

$$v_z \frac{\partial v_z}{\partial z} + v_r \frac{\partial v_z}{\partial r} = \nu \left(\frac{\partial^2 v_z}{\partial z^2} + \frac{\partial^2 v_z}{\partial r^2} + \frac{1}{r} \frac{\partial v_z}{\partial r} \right) + g,$$

$$v_z \frac{\partial v_r}{\partial z} + v_r \frac{\partial v_r}{\partial r} = \nu \left(\frac{\partial^2 v_r}{\partial z^2} + \frac{\partial^2 v_r}{\partial r^2} + \frac{1}{r} \frac{\partial v_r}{\partial r} - \frac{v_r}{r^2} \right);$$

$$z = 0, \quad v_z = v_z^0, \quad \frac{\partial v_z}{\partial z} = 0, \quad v_r = 0, \quad \frac{\partial v_r}{\partial z} = 0;$$

$$r = 0, \quad \frac{\partial v_z}{\partial r} = 0, \quad \frac{\partial v_r}{\partial r} = 0; \quad r = r_0 - \delta(z), \quad v_z = v_0(z), \quad v_r = \frac{r_0 v_0(z)}{r_0 - \delta(z)} \frac{d\delta}{dz};$$

$$\delta(z) = \frac{z}{a + bz}, \quad \frac{d\delta}{dz} = \frac{a}{(a + bz)^2}, \quad v_0(z) = \frac{z}{\alpha + \beta z}.$$

$$(1.25)$$

9.3   Liquid Layer Hydrodynamics

The wall flow in the column is described by the equations:

$$w_z \frac{\partial w_z}{\partial z} + w_r \frac{\partial w_z}{\partial r} = \nu \left(\frac{\partial^2 w_z}{\partial z^2} + \frac{\partial^2 w_z}{\partial r^2} + \frac{1}{r} \frac{\partial w_z}{\partial r} \right) + g,$$

$$w_z \frac{\partial w_r}{\partial z} + w_r \frac{\partial w_r}{\partial r} = \nu \left(\frac{\partial^2 w_r}{\partial z^2} + \frac{\partial^2 w_r}{\partial r^2} + \frac{1}{r} \frac{\partial w_r}{\partial r} - \frac{w_r}{r^2} \right);$$

$$z = 0, \quad w_z = 0, \quad \frac{\partial w_z}{\partial z} = 0, \quad w_r = 0, \quad \frac{\partial w_r}{\partial z} = 0;$$

$$r = r_0, \quad w_z = 0, \quad w_r = 0; \quad r = r_0 - \delta(z), \quad w_z = v_0(z), \quad w_r = v_r = \frac{r_0 v_0(z)}{r_0 - \delta(z)} \frac{d\delta}{dz};$$

$$\delta(z) = \frac{z}{a + bz}, \quad \frac{d\delta}{dz} = \frac{a}{(a + bz)^2}, \quad v_0(z) = \frac{z}{\alpha + \beta z}.$$

$$(1.26)$$

9.3.1   Parameters Identification

The comparison of (1.25) and (1.26) shows that there are a common boundary condition at an unknown boundary $\delta(z)$:

$$r = r_0 - \delta(z), \quad w_z = v_z = v_0(z), \qquad (1.27)$$

where the surface velocity of the wall flow $v_0(z)$ is unknown too.

The flow rate of the liquid flowing on the wall, per unit of column circumference $Q(z)$ in (1.11), which has passed from the bulk of the liquid in the column, depends on the functions $\delta(z)$, $v_0(z)$ and must be determined by the equations:

$$Q(z) = \frac{z}{m_1 + m_2 z} = \int_{r_0 - \delta(z)}^{r_0} w_z dr = \int_0^{r_0 - \delta(z)} \left[v_z^0 - v_z \right] dr, \quad v_z^0 = \frac{Q_L}{\varepsilon 2\pi r_0^2}, \quad (1.28)$$

i.e., the flow rate of the wall flow is equal to the difference in the flow rates of the liquid in the bulk of the column in presence and absence of a wall flow. The conditions (1.28) and the solution of the system of eqs. (1.25 and 1.26) permit to be obtained the parameters (a, b, α, β), using a suitable algorithm [7].

9.3.2 Gas Phase Hydrodynamics

The hydrodynamics of the gas phase will be represented in a cylindrical coordinate system (r, z_0), $z_0 = l - z$, where the axial coordinate is directed back to the axial coordinate of the liquid phase. The movement of the gas is the result of the pressure gradient along the height of the column (hydraulic resistance), which depends on the packings and is determined experimentally. The axial u_z and radial u_r components of velocity in the gas phase and pressure (per unit volume) p satisfy the Navier–Stokes equations:

$$u_z \frac{\partial u_z}{\partial z_0} + u_r \frac{\partial u_z}{\partial r} = -\frac{1}{\rho} \frac{\partial p}{\partial z_0} + v \left(\frac{\partial^2 u_z}{\partial z_0^2} + \frac{\partial^2 u_z}{\partial r^2} + \frac{1}{r} \frac{\partial u_z}{\partial r} \right),$$

$$u_z \frac{\partial u_r}{\partial z_0} + u_r \frac{\partial u_r}{\partial r} = -\frac{1}{\rho} \frac{\partial p}{\partial r} + v \left(\frac{\partial^2 u_r}{\partial z_0^2} + \frac{\partial^2 u_r}{\partial r^2} + \frac{1}{r} \frac{\partial u_r}{\partial r} - \frac{u_r}{r^2} \right),$$

$$\frac{\partial u_z}{\partial z_0} + \frac{\partial u_r}{\partial r} + \frac{u_r}{r} = 0; \tag{1.29}$$

$$z_0 = 0, \quad u_z = u_z^0, \quad \frac{\partial u_z}{\partial z_0} = 0, \quad u_r = 0, \quad \frac{\partial u_r}{\partial z_0} = 0, \quad p = p^0;$$

$$r = 0, \quad \frac{\partial u_z}{\partial r} = 0, \quad \frac{\partial u_r}{\partial r} = 0, \quad p = p(0, z);$$

$$r = r_0 - \frac{l - z_0}{a + b(l - z_0)}, \quad u_z = -v_0(l - z_0), \quad u_r = 0.$$

The pressure in the gas phase $p(r, z_0)$ is possible to be presented in (1.29) as

$$\frac{\partial p}{\partial r} \equiv 0, \quad \frac{\partial p}{\partial z_0} = H = \frac{p^0 - p(0, z_0)}{z_0}, \tag{1.30}$$

where H is the packing pressure drop, i.e., the pressure drop through a packing layer of a thickness of 1 meter at a given gas velocity.

10   Modeling of Processes with Unknown Mechanism

There are complex processes whose mechanism is unknown. A typical example of this is complex chemical reactions, the rate of which depends on the concentrations of several substances, but the simple chemical reactions and the relationships between them are unknown.

The kinetics of processes with an unknown mechanism can be modeled on the basis of the axiom, according to which "The mathematical structure of the quantitative description of real processes does not depend on the measuring system of the quantities involved in them." On the basis of this axiom, Guchmann's theorem can be proved [8]—"If mathematical structure is invariant with respect to similar transformations, it is possible to be presented as power functions complex," i.e., "Mathematical structure of the quantitative description of real systems is possible to be presented as power functions complex," because the mathematical structure which is invariant with respect to similar transformations is mathematical structure of real processes which does not depend on the measuring system.

The kinetics of the chemical engineering processes depends on a set of variables. If the velocity of these processes is denoted by y and the values of these variables are $x_1, \ldots, x_n$, the equation of the kinetic model will have the form:

$$y = f(x_1, \ldots, x_n). \tag{1.31}$$

This function is a mathematical structure that is retained when changed the measurement system of the variable, i.e., this mathematical structure is invariant with respect to similar transformations [7]:

$$\bar{x}_i = k_i x_i, \quad i = 1, \ldots, n, \tag{1.32}$$

i.e., f is a homogeneous function:

$$ky = f(k_1 x_1, \ldots, k_n x_n) = \phi(k_1, \ldots, k_n).f(x_1, \ldots, x_n), \quad k$$
$$= \phi(k_1, \ldots, k_n). \tag{1.33}$$

A short recording of (1.33) is:

$$f[\bar{x}_i] = \phi[k_i] \, f[x_i]. \tag{1.34}$$

The problem consists in finding a function f that satisfies Eq. (1.34). A differentiation of Eq. (1.34) concerning k_1 leads to:

$$\frac{\partial f[\bar{x}_i]}{\partial k_1} = \frac{\partial \phi}{\partial k_1} \, f(x_i). \tag{1.35}$$

On the other hand

$$\frac{\partial f[\bar{x}_i]}{\partial k_1} = \frac{\partial f[\bar{x}_i]}{\partial \bar{x}_1} \frac{\partial \bar{x}_1}{\partial k_1} = \frac{\partial f[\bar{x}_i]}{\partial \bar{x}_1} \, x_1. \tag{1.36}$$

From (1.35 and 1.36) follows

$$\frac{\partial f[\bar{x}_i]}{\partial \bar{x}_1} \, x_1 = \alpha_1 \, f[x_i], \tag{1.37}$$

where

$$\alpha_1 = \left(\frac{\partial \phi}{\partial k_1} \right)_{k_i=1}. \tag{1.38}$$

The Eq. (1.37) is valid for different values of k_i including $k_i = 1$ ($i = 1, \ldots, n$). As a result $\bar{x}_i = x_i$, $i = 1, \ldots, n$ and from (1.37) follows

$$\frac{1}{f} \frac{\partial f}{\partial x_1} = \frac{\alpha_1}{x_1}, \tag{1.39}$$

i.e.,

$$f = c_1 x_1^{\alpha_1}. \tag{1.40}$$

When the above operations are repeated for $x_2, \ldots, x_n$, the homogenous function f assumes the form:

$$f = k x_1^{\alpha_1}, \ldots, x_n^{\alpha_n}, \tag{1.41}$$

i.e., the function f is homogenous if it represents a power functions complex and as a result is invariant with respect to similarity (metric) transformations. The parameters $k, \alpha_1, \ldots, \alpha_n$ are determined by experimental data of the industrial process velocity.

The power functions complex (1.41) is used in the similarity criterion models equation in the similarity theory [7], where x_i, $i = 1, \ldots, n$ are similarity criterions. The information about the process mechanism permits to be obtained dimensionless combinations of the model parameters (similarity criteria), which represent the ratio of the efficiency of two processes. For example [1]:

$$\text{Fo} = \frac{Dl}{u^0 r_0^2}, \quad \text{Da} = \frac{kl}{u^0},$$ (1.42)

are the Fourier and Damkohler numbers and represent the ratio of the efficiency of the convective and diffusion mass transfer (Fo) and convective mass transfer and chemical reaction rate (Da). In (1.42), D is diffusivity, r_0, l—linear characteristic scales, u^0—velocity characteristic scale.

The power functions complex (1.41) is used by the dimension analysis [1], but the dimension criteria do not represent the ratio of the efficiency of two processes, because the method is used in absence of the processes mechanism information.

11 Conclusions

In the paper is presented a theoretical analysis of the methods for chemical engineering processes modeling. The methods for modeling specific processes may be different, but in all cases they must bring the mathematical description closer to the real process by using appropriate experimental data.

The role of the kinetics of industrial apparatuses for solving the problems of optimal design and control is analyzed. The thermodynamic, hydrodynamic, and Boltzmann approximations for the mathematical description of the kinetics of industrial apparatuses are described.

They are presented the cases of co-current absorption column without packings, counter-current absorption column with random packings, and modeling of processes with unknown mechanism.

Acknowledgments This work is supported by Project of Fundamental Scientific Research 19-58-18004, conducted by RFBR and the National Science Fund of Bulgaria under contract No KP 06 RUSIA-3 from 27 Sep. 2019, "Modeling, simulation and experimental investigations of the interphase mass transfer and separation in distillation, absorption, adsorption and catalytic processes in industrial column apparatuses."

References

1. C. Boyadjiev, *Theoretical Chemical Engineering. Modeling and Simulation* (Springer, Berlin, 2010)
2. J. Keizer, *Statistical Thermodynamics of Nonequilibrium Processes* (Springer, New York, 1987)
3. C. Boyadjiev, Reaction kinetics in chemical engineering. Bulgarian Chemical Communications **50**, 112–119 (2018)
4. C. Boyadjiev, Some Thoughts on Logic and Intuition in Science and Chemical Engineering. Open Access Library Journal **1**(6), 1–5 (2014)

5. C. Boyadjiev, M. Doichinova, B. Boyadjiev, P. Popova-Krumova, *Modeling of Column Apparatus Processes* (Springer, Berlin, 2016)
6. C. Boyadjiev, M. Doichinova, B. Boyadjiev, P. Popova-Krumova, *Modeling of Column Apparatus Processes*, 2nd edn. (Springer, Berlin, 2018)
7. B. Boyadjiev, C. Boyadjiev, D. Dzhonova, A.N. Pavlenko, On the hydrodynamics of gas and liquid counter-current flows in packed columns. Private Communication
8. А. Гухман. Введение в теорию подобия, Изд. „Высшая школа", Москва (A. Gukhman, Introduction to the theory of similarity, Ed. „Higher School", Moscow), 1973

Chapter 2
Modeling of Ethanol Fermentation from Low-Grade Raw Materials, Including Cellulose and Hemicellulose in a Two-Step Bioreactor

V. Beschkov, P. Popova-Krumova, D. Yankov, G. Naydenova, and I. Valchev

Abstract The successful ethanol fermentation from lignocellulosic substrates is impeded by the fact that cellulose and hemicellulose must be digested by different strains, hence full substrate utilization is not possible by one microbial strain in one process. That is why these two polysaccharides should be treated separately by different microbial strains. The simplest way to accomplish these processes is to use bioreactor with separated compartments where cellulose and hemicellulose are fermented by different strains: *Saccharomyces cerevisiae* for hexoses in cellulose and *Pichia stipitis* for pentoses in hemicellulose. In the present work such a two-step process is modeled for a continuous operation. As substrates glucose (a hexose) and xylose (a pentose) resulting of acid pre-treatment of the lignocellulose are used. The addition of the produced ethanol in the first compartment to the second one as starting substrate and its inhibition impact are taken into account. The dilution rate, the initial substrate concentration, and the inhibition effects are considered too.

Mathematical models describing these fermentation processes are composed and used for kinetic parameter estimation from own experimental data. Both product and substrate inhibitions are taken into account. It is demonstrated that the simultaneous glucose and xylose fermentation to ethanol by the strain *Pichia stipitis* is successfully described by the composed model. The conversion from monosaccharides to ethanol takes place with yield of 70%.

Key words Ethanol fermentation · Cellulose · Hemicellulose · Two-step bioreactor · Modeling · Parameter evaluation

V. Beschkov (✉) · P. Popova-Krumova · D. Yankov · G. Naydenova
Institute of Chemical Engineering, Bulgarian Academy of Sciences, Sofia, Bulgaria

I. Valchev
Department for Pulp and Paper, University of Chemical Technology and Metallurgy, Sofia, Bulgaria

Notations

Dr	Dilution rate, h^{-1}
ki	Substrate inhibition constant, $g.dm^{-3}$
kp	Inhibition constant, $g.dm^{-3}$
k_S	Saturation constant, $g.dm^{-3}$
n	Power value in Eq. (2.1)
P	Product (ethanol) concentration, $g.dm^{-3}$
Q	Substrate feeding flow rate for continuous process, $dm^3 \, h^{-1}$
S	Substrate concentration, $g.dm^{-3}$
t	Time, h
V	Reactor volume, dm^3
X	Biomass (microbial cell) concentration, $g.dm^{-3}$
$Y_{P/S}$	Product yield coefficient, g/g
$Y_{X/S}$	Biomass yield coefficient, g/g

Greek Symbols

α	Kinetic rate constant for microbial growth-associated ethanol production, $[-]$
β	Kinetic rate constant for non-growth-associated ethanol production, h^{-1}
γ	Rate constant for ethanol decay, h^{-1}
μ	Specific microbial growth rate constant, h^{-1}

Indices

1	Refers to processes in reactor 1 and to glucose as substrate with *S.cerevisiae*
2	Refers to processes in reactor 2 and to xylose as substrates with *P. stipitis*
crit	Refers to critical inhibition value in Eq. (2.1)
i	Refers to substrate inhibition constant in Eq. (2.1)
max	Refers to maximum specific growth rate
P	Refers to product inhibition constant in Eqs. (2.2 and 2.7)
S	Refers to saturation constant in microbial growth equations, Eqs. (2.1, 2.2, 2.7)

1   Introduction

The extended practical application of ethanol as fuel or fuel additive was hindered due to economic and social reasons. The main problem consisted in the extensive use of cereals as renewable substrate for ethanol fermentation. This fact leads to substantial increase of the cereal prices on the world market making the poor countries

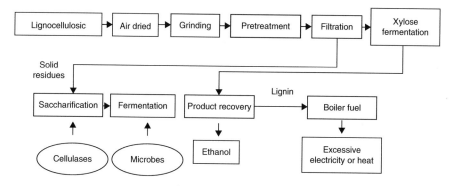

Scheme 2.1 Schematic overview of biomass-to-ethanol conversion process [5].

poorer. That is why attention was paid to the so-called "second generation" raw materials, i.e., lignocellulosic waste from agriculture. The lignocellulosic materials are mainly composed of cellulose, hemicellulose, lignin, and other minor components. The cellulose-based bioethanol production includes the following steps [1].

The main problem in this direction is the necessity of saccharification preliminary to the fermentation step either by chemical or enzyme processes. The fermentative ethanol production from lignocellulose feedstock consists in several consecutive steps, cf. Scheme 2.1.

1. Pre-treatment of the raw material to break down the hemicellulose and lignin structures in order to improve the substrate digestibility [2].
2. Saccharification of the raw material. The traditional ones are starch containing cereal crops being hydrolyzed by chemicals (e.g., sulfuric acid [3, 4]) or amylolytic enzymes (alpha-amylase amyloglucosidase) to mono- and oligosaccharides. Lignocellulose is saccharified by cellulases [3].
3. Removal or separation of lignin from the hydrolyzate.
4. Fermentation to yield some 10–12% vol. of ethanol.
5. Ethanol concentration up to 96% vol. by distillation.
6. Production of absolute ethanol (more than 99% ethanol content) by molecular sieve applications.

Hemicellulose is a complex, heterogeneous mixture of sugars and sugar derivatives that form a highly branched network. The monomers that comprise hemicellulose are hexoses (glucose, galactose, and mannose) and pentoses (arabinose and xylose). Those monosaccharides are released after hydrolysis.

Lignin and associated phenolic acids, although present in relatively small concentrations, play an important role in cell wall degradation. The chemical structure of lignin is very complicated [6], as it is a three-dimensional cross-linked aromatic polymer made up from phenylpropane units. No single established structural scheme for lignin has been established so far.

The fermentation step for hexoses as substrate is made by yeasts, like *S. cerevisiae* and bacteria, e.g., *Z. mobilis* [7]. However, these microbes are not capable to ferment pentoses, like arabinose and xylose, which are products of hemicellulose pre-treatment and saccharification. Other microbes like *Pichia stipitis* [8–10] are needed. That is why, the complete conversion to ethanol needs either separate fermentation of hexoses and pentoses in different fermenters or to do it in one step in one bioreactor. The latter approach poses some difficulties due to formation of various intermediates which might inhibit the target processes, the competitive growth and metabolism of the different microbes, etc.

In the present work, own experimental data on the parallel fermentation of glucose (S1) and xylose (S2) accomplished by yeasts *Saccharomyces cerevisiae* (I) and *Pichia stipitis* (II) in two individual bioreactors are presented. Mathematical models encountering two types of product inhibition will be applied to handle the experimental data.

2 Mathematical MODELING

There are some efforts for modeling the ethanol fermentation from glucose [7, 11], as well as the simultaneous ethanol production from glucose and xylose by single microbial strain, usually modified by genetic procedures [7, 12–14], or by consortium of two strains, separately converting hexoses and pentoses to ethanol [15]. In [15], product inhibition is described by Monod-Yerusalimskii growth kinetics with no substrate inhibition involved. Only microbial growth-associated ethanol production is presumed in [11, 15].

The simple ethanol fermentation from glucose has been modeled by Oliveira et al. [11], cf. Eqs (2.1).

$$
\begin{aligned}
\frac{dX}{dt} &= \mu X, \mu = \mu_{max} \frac{S}{k_S + S + \frac{S^2}{k_i}} \left(1 - \frac{P}{P_{crit}}\right)^n, 0 \leq n \leq 1 \\
\frac{dP}{dt} &= \alpha \mu X \\
\frac{dS}{dt} &= \frac{dP}{dt} / Y_{P/S}
\end{aligned}
\tag{2.1}
$$

In this case growth-associated fermentation is considered with Levenspiel-type product inhibition. The yield coefficient $Y_{P/S}$ is given by the stoichiometric ratio ($Y_{P/S} = 0.511$ g/g) or by other empirical considerations, for example $Y_{P/S} = 0.40$ g/g, as assumed in [11].

There are different attempts for the modeling of ethanol fermentation, applying various kinetic equations for microbial growth [7, 11–13]. Basically two types of kinetic models are applied: the Levenspiel's model (2.1) and the one of Monod-Yerusalimskii (2.2), taking into account the product (P) inhibition, i.e., ethanol:

$$\frac{\partial X}{\partial t} = \mu X, \quad \mu = \mu_{\max} \frac{k_P}{k_P + P} \frac{S}{k_S + S} \tag{2.2}$$

In the first case (Eq.2.1), the substrate inhibition of microbial growth is considered by the Haldane's model. The substrate consumption and the product formation are associated with the microbial growth. In this paper we shall consider product conversion associated both with microbial growth and with the culture in its stationary state. For modeling of these processes, we present by the following system of ordinary differential equations with the appropriate initial conditions.

$$\frac{\partial S}{\partial t} = -\frac{\alpha}{Y_{X/S}} \mu X - \beta X \tag{2.3}$$

$$\frac{dP}{dt} = \frac{\cdot}{Y_{X/S}} X + X - \gamma P \tag{2.4}$$

$$t = 0, X = X_0, S = S_0, P_0 = 0 \tag{2.5}$$

The model (2.3–2.5) assumes both growth-associated (by the term $\alpha\mu X$) and non-growth-associated ethanol production by the term βX and further consecutive conversion of ethanol (as intermediate) into final products like acetaldehyde and acetic acid by the term γX. This model, Eqs. (2.2–2.5) comprises the model (2.1) as particular case when $\beta = 0$ and $\gamma = 0$.

In all these cases we presume that only already hydrolyzed substrates including low molecular oligosaccharides and monosaccharides (i.e., glucose and xylose) are subjected to fermentation.

2.1 One-Step Fermentation

This process is related to batch or continuous process where only one substrate (i.e., glucose) is digested and fermented to ethanol. We shall consider the case of product inhibition according to the Monod-Yerusalimskii model, cf. Eq. (2.2). In this case the system of ordinary differential equations comprises Eqs. (2.6) with initial conditions, Eq. (2.5).

2.2 Two-Step Fermentation

Things become more complicated at the two-step fermentation because of the presence of hydrolyzates and monomers from cellulose and hemicellulose. There are two approaches to carry out these processes: in one-step process involving two separate strains digesting glucose and xylose simultaneously in one reactor and

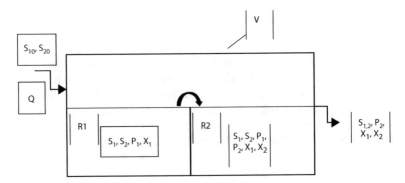

Fig. 2.1 A sketch of the two-step fermentation in bioreactor with separated compartments

another one, where the two processes are separated spatially and take place in two consecutive bioreactors with two separate strains, cf. Figure 2.1.

The alcohol fermentation of hexoses (glucose in particular) is carried out in the first compartment R1 and the pentoses (xylose, namely)—in the second one, R2. The residual glucose from compartment R1 enters the compartment R2 and can be subjected to fermentation by the second strain. It is important that the ethanol, being product in R1 is supplied to reactor R2 where the second strain suitable for xylose fermentation is inoculated. Note that the ethanol is inhibitor for the fermentation in the second reactor. The continuous mode of the two-step process helps to minimize the adverse effect of product inhibition.

In the present study we shall consider glucose and xylose as substrates representative for hexoses and pentoses. The continuous two-step fermentation will be described by the following system ordinary differential equations:

$$D_r X_1 = \mu_1 X_1$$
$$D_r \left(S_1^0 - S_1 \right) = -\frac{\alpha_1}{Y_{X/S}} \mu_1 X_1 - \beta_1 X_1$$
$$D_r \left(P_1^0 - P_1 \right) = Y_{P/S1} \frac{\alpha_1}{Y_{X/S1}} \mu_1 X_1 + \beta_1 X_1$$
$$D_r X_2 = \mu_2 X_2 \tag{2.6}$$
$$D_r \left(S_2^0 - S_2 \right) = -\frac{\alpha_2}{Y_{X/S2}} \mu_2 X_2 - \beta_2 X_2$$
$$D_r \left(P_2^0 - P_2 \right) = Y_{P/S2} \frac{\alpha_2}{Y_{X/S2}} \mu_2 X_2 + \beta_2 X_2$$
$$D_r = Q/V$$

Here S_1 and S_2 denote the two substrates—glucose (S_1) and xylose (S_2), and ethanol is denoted by P with indices "1" and "2" for the first and the second compartment, respectively. The microbial cell concentrations are denoted by X_1 for *Saccharomyces cerevisiae* and for *Pichia stipitis*—X_2. The dilution rate D_r is the ratio between the feeding flow rate Q and the reactor volume V.

For the case when both glucose and xylose are simultaneously fermented, the mathematical model with the associated initial conditions is the following:

$$\frac{\partial X_2}{\partial t} = \mu_2 X_2 - D_r X_2, \quad \mu_2 = \mu_{\max 2} \frac{k_{P_2}}{k_{P_2} + P_2} \frac{S_1}{k_{S_1} + S_1} \frac{S_2}{k_{S_2} + S_2}$$

$$\frac{\partial S_1}{\partial t} = D_r \left(S_1^0 - S_1 \right) - \frac{\alpha_1}{Y_{X_2/S_1}} \mu_2 X_2 - \beta_1 X_2 \tag{2.7}$$

$$\frac{\partial S_2}{\partial t} = D_r \left(S_2^0 - S_2 \right) - \frac{\alpha_2}{Y_{X_2/S_2}} \mu_2 X_2 - \beta_2 X_2$$

$$\frac{dP_2}{dt} = D_r \left(P_2^0 - P_2 \right) + Y_{P_2/S_2} \left\{ \left[\frac{\alpha_1}{Y_{X_2/S_1}} + \frac{\alpha_2}{Y_{X_2/S_2}} \right] X_2 + (\beta_1 + \beta_2) X_2 \right\} - \gamma P_2$$

$$t = 0, \quad X_2 = X_2^0, \quad S_1 = S_1^0, \quad S_2 = S_2^0, \quad P_2 = P_2^0$$

The feeding flow rate is denoted by Q and V is the total volume of the bioreactor. The ratio Q/V is the dilution rate D_r. At $D_r \to 0$, batch mode processes take place.

The systems (2.1), (2.6), and (2.7) were solved by the software package MATLAB 2013A, and the kinetic constants were determined from experimental data after optimization procedure by the least squares method. Both one-step and two-step fermentation processes are available.

The results of the mathematical modeling were coupled to own experimental data for verification and model parameter estimation minimizing the sum of the squares of the computed and the experimental values for the substrates (glucose and/or xylose) and the product: ethanol and biomass:

$$Q = \frac{1}{N} \sum_{i=1}^{N} \left[X(t_i) - X_{\exp}(t_i) \right]^2 + \frac{1}{N} \sum_{i=1}^{N} \left[S(t_i) - S_{\exp}(t_i) \right]^2 + \frac{1}{N}$$

$$\times \sum_{i=1}^{N} \left[P(t_i) - P_{\exp}(t_i) \right]^2, \tag{2.8}$$

where $t_i (i = 1, \ldots, N)$ is the numbers of sampling during the experiment.

The experimental conditions are described below.

3 Experimental

The following experiments of alcohol fermentation using two different substrates were carried out.

3.1 Alcohol Fermentation of Acid Hydrolyzate of Lignocellulosic Materials

Substrate was supplied by the Bulgarian company Svilosa SA. In this study the residual hexoses (glucose, etc.) and the oligosaccharides were subjected to alcohol fermentation by the yeast *Saccharomyces cerevisiae* after acid pre-treatment. The strain *Saccharomyces cerevisiae Meyen and Hansen 1883 (NBIMCC 541)* was supplied by the National Bank of Industrial Microorganisms and Cell Cultures (Bulgaria). It is equal to the strains ATCC 9763, CBS 2978, DSMZ 1333, NCYC 87.

The yeast culture was maintained on medium № 285 (of the National Bank of Industrial Microorganisms and Cell Cultures, Bulgaria) with a content, containing the following components, g/l: yeast extract, 3; malt extract, 3; peptone, 5; glucose, 10. The components were dissolved in 1 liter of distilled water. The acidity was adjusted to pH 6.0 by NaOH. The medium was sterilized for 20 min at 121 °C.

The strain was cultivated at 25 °C, for 18–24 h in the above liquid medium. Afterwards the experiments were carried out at 25 and 30 °C with and without addition of glucose to the broth. Experiments with *Saccharomyces cerevisiae* were performed on medium 285 with added 5 g/l glucose (reactor R1) and hydrolyzate provided by Svilosa SA as a liquid phase. The experiments were performed at 25 °C, aerobically, static, for 72 h. Samples were taken for reducing sugars, biomass, and for glucose, xylose and ethanol. The experiments were carried out in shaking flasks in a rotary shaker (New Brunswick Scientific, NJ, USA) at 200 rpm or under static conditions (in reactors R1 and R2).

The hydrolyzate contains also pentoses as product of hemicellulose hydrolysis, like D-xylose. Therefore, another strains capable to digest pentoses are required. Usually they are from the *Pichia* genus.

This process was carried out in an individual reactor R2 using the yeast *Schefferomyces stipitis (Pichia stipitis) CBS 7124*. The strain *Schefferomyces stipitis (Pichia stipitis) CBS 7124* was supplied by the National Bank of Industrial Microorganisms and Cell Cultures (Bulgaria).

The strain was cultivated at 25 °C, for 18–24 h in a liquid medium GPYA (Glucose-peptone-yeast extract) (of the National Bank of Industrial Microorganisms and Cell Cultures, Bulgaria), containing the following components, g/l: yeast extract, 5; peptone, 5; glucose, 40. The components were dissolved in 1 liter of distilled water.

The experiments, for ethanol production, from cellulose hydrolyzate (supplied by the Bulgarian company Sviloza SA), by *Pichia stipitis CBS 7124*, were performed in 3 reactors, 60 ml each. Fermentation medium (GPYA) and hydrolyzate were in different ratio (hydrolyzate: GRYA; 1:1; 1:3; 1:4). The components of GPYA (without glucose) were added to the hydrolyzate. The hydrolysate contained also 5 g/l xylose. The experiment was carried out aerobically at 27 °C, under static conditions, for 173 h. The content of the lignocellulose substrate after acid hydrolysis is given in Table 2.1.

Table 2.1 Content of the lignocellulose hydrolyzate supplied by Sviloza SA

Component	Concentration, g dm^{-3}
D-cellobiose	0.149
D(+)-glucose	0.879
D(+)-xylose	3.870
D(+)-manose	0.091
Hydroxymethylfurfural	0.056
Furfural	0.007

This substrate was treated in parallel by *S. cerevisiae* in reactor R1 and by *P. stipitis* in reactor R2 for xylose digestion. The two reactor system comprised glucose and xylose fermentation where the fermentation process by *S. cerevisiae* is taking place in reactor R1 and the fermentation by *P. stipitis* is accomplished in reactor R2. Note that *P. stipitis* is capable to convert glucose to ethanol too.

Sample from both reactors were taken regularly and they were analyzed for sugars and ethanol. In parallel, the biomass concentration was monitored by the sample turbidity.

3.2 Analyses

The different sugar content as well as the ethanol concentration were determined by HPLC under the following conditions. The HPLC system consisted of a pump Smartline S-100, Knauer, refractometric detector—Perkin Elmer LC-25RI, column Aminex HPX-87H, Biorad, 300x7, 8 mm and specialized software Knauer. As a mobile phase, 0.01 N H_2SO_4 was used, at a flow rate 0.6 ml/min. The biomass concentration was determined by optical density measurements at 620 nm with a spectrophotometer VWR UV-1600PC. The total reducing sugars concentration were determined titrimetrically by the Bertrand method.

4 Results and Discussion

4.1 Results of the Mathematical Modeling

Some model results on ethanol fermentation with inhibition described by the Monod-Yerusalimskii model, Eq. (2.2) with the system (2.6) at different dilution rates are shown in Fig. 2.2.

The effect of the residence time, i.e., the dilution rate on the fermentation process is shown. The limiting case when Dr. → 0 corresponds to the batch process with strong variation of the concentrations in time. The results presented in Fig. 2.2 show that at Dr. < 0.001 the system behaves as under batch conditions.

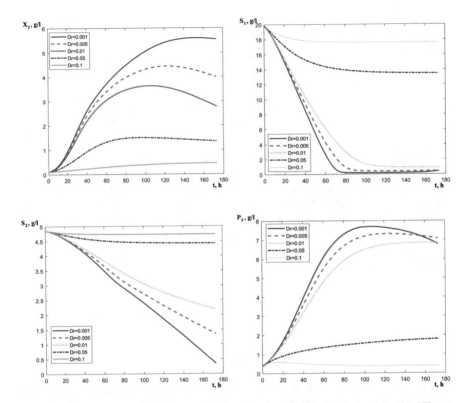

Fig. 2.2 Results of mathematical modeling for ethanol production by *S. cerevisiae* for different dilution rates. Biomass (X_1), substrate (S_1 and S_2) and product (P_2) time profiles for different dilution rates. Kinetic parameters are estimated from Experiment 3, Table 2.2

4.2   Results of Parameter Estimation from Experimental Data by the Mathematical Models

Ethanol from cellulose hydrolyzate by S. cerevisiae. The results obtained by the mathematical modeling of ethanol production from glucose, cf. Eqs. (2.3–2.5), at product inhibition described by the model of Monod-Yerusalimskii, Eq. (2.2) by *S. cerevisiae* are compared to own experimental data. They are shown in Fig. 2.3. One reactor R1 is considered. The estimated model parameters are shown in Table 2.2.

The evaluated parameters in the model (2.3–2.5) for different experiments are shown in Table 2.2. The yield coefficient $Y_{P/S} = 0.511$ is the stoichiometric one for glucose (or xylose) conversion to ethanol:

The presented model predicts the further conversion of ethanol to acetaldehyde and acetic acid. The comparison shows that the estimated parameters for two experiments are in good agreement, particularly for the maximum specific growth rate μ_{max}, and the production rate constant β for the stationary phase of the culture.

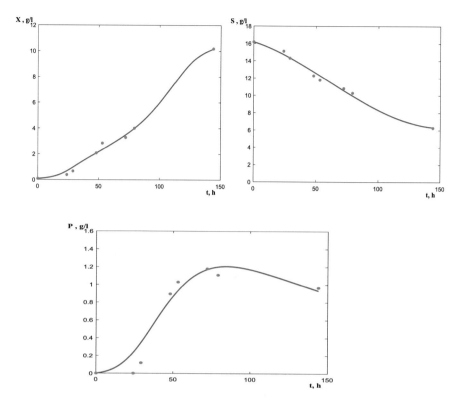

Fig. 2.3 Data for glucose fermentation by *S. cerevisiae*. Concentration profiles for biomass (*X*), glucose (*S*), and ethanol (*P*) compared to the experimental data after parameter evaluation. $X_0 = 0.083$; $S_0 = 16.1693$; $P_0 = 0$.; $\mu_{max} = 0.2029$ h^{-1}; $k_p = 0.4537$; $k_s = 15{,}023$; $\alpha/Y_{X/S} = 1.019$; $Y_{P/S} = 0.511$; $\beta = 0.073$ h^{-1}; $\gamma = 0.058$ h^{-1}. Concentrations in g/l.

Table 2.2 Model parameters for fermentation of glucose by *S. cerevisiae* in reactor R1. Substrate—cellulose hydrolyzate. Dr. $= 0.001$ h^{-1}. Concentrations are in g dm^{-3}. Experiment 1: $X_0 = 0.106$; $S_0 = 16.5$; $P_0 = 0$. Experiment 2: $X_0 = 0.083$; $S_0 = 16.2$; $P_0 = 0$

Parameters	Experiment 1	Experiment 2
μ_{max}, h^{-1}	0.246	0.2029
k_P, g dm^{-3}	0.3247	0.4537
k_S, g dm^{-3}	1.3707	1.5023
$\alpha/Y_{X/S}$, [−]	1.0784	1.0919
β, h^{-1}	0.0071	0.0073
γ, h^{-1}	0.0065	0.0058

The values of the rate constant β for the non-growth-associated ethanol production show moderate contribution compared to the growth-associated one. The other model parameters, i.e., the saturation constant k_S, the product inhibition constant k_P differ, but they are of the same order of magnitude. That is why the proposed

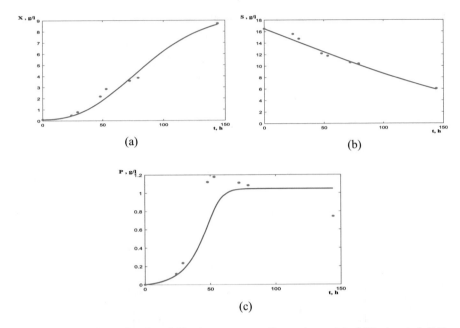

Fig. 2.4 Comparison of evaluated kinetic curves according to the model of Oliveira et al. [11], cf. Table 2.4, with experimental data, Experiment 2. Biomass (*S. cerevisiae*), substrate, and product time profiles for Dr. = 0.001 h^{-1}; (a) biomass; (b) glucose; (c) ethanol

model fits well our experimental data and it can be used for predictions on the fermentation operation for making decisions on the process conduction.

Comparison with the model of Oliveira **et al.** [11]. We have tried to handle our experimental data by the model of Oliveira et al. [11] where the ethanol production by glucose was considered as growth associated only, cf. Eqs (2.1). The results are demonstrated in Fig. 2.4 and Table 2.3. A comparison of the evaluated parameters by the present model, i.e., Eqs. (2.6), Table 2.2 and the values by the model of Oliveira et al. [11] was made in Table 2.3. The main growth kinetic parameters, i.e., the maximum specific growth rate μ_{max} and the saturation constants k_S estimated by these two models are comparable. Although the curves of the Oliveira model fit well the experimental data for biomass and glucose, there is large discrepancy for the results for ethanol because the model [11] does not take into account the further decay of ethanol because of its further conversion, cf. Fig. 2.4. It is obvious that our model, Eqs. (2.6) fits better the experimental results, cf. Fig. 2.3 because it takes into account the ethanol conversion to other products.

Ethanol from cellulose hydrolyzate by P. stipitis. The values of the estimated parameters for simultaneous glucose and xylose fermentation by *P. stipitis* by the model (2.7) are shown in Table 2.4. Besides the saturation constants, rate and yield coefficients k_{S1}, $\alpha_1/Y_{X2/S1}$, $\alpha_2/Y_{X2/S2}$, β_2, the scatter of the determined parameters are relatively low, i.e., below 10%. It is an indication for the validity of the used model. The comparison of the model and the experimental data is shown in Fig. 2.5. There is

Table 2.3 Parameters for glucose to ethanol fermentation estimated by the model of Oliveira et al. [11], Eqs. (2.1). Own experimental data obtained by *S. cerevisiae*. Concentrations are as they are in Table 2.2

Parameter	Experiment 1		Experiment 2	
	Oliveira et al. [11]	Present model, Table 2.2	Oliveira et al. [11]	Present model, Table 2.2
μ_{max}, h^{-1}	0.2357	0.246	0.291	0.2029
k_S, g dm^{-3}	1.1839	1.3707	1.202	1.5023
k_I, g dm^{-3}	0.4042	n.a.	0.499	n.a.
k_P, g dm^{-3}	n.a.	0.3247	n.a.	0.4537
α, [−]	1.2825	1.0784	1.5875	1.0919
P_{max}	63.42	n.a.	64.583	n.a.
n	0.715	n.a.	0.758	n.a.
β, h^{-1}	n.a.	0.0071	n.a.	0.0073
γ, h^{-1}	n.a.	0.0065	n.a.	0.0058

Table 2.4 Model parameters for simultaneous fermentation of glucose and xylose by *P. stipitis* in reactor R2 and Dr. = 0.001 h^{-1}

Parameter	Experiment 1	Experiment 2	Experiment 3	Average	Scatter, %
μ_{max} h^{-1}	0.6756	0.6982	0.7869	0.7202	8
k_P g.dm^{-3}	0.2621	0.2951	0.2746	0.2772	6
k_{S1} g.dm^{-3}	0.8648	0.7001	0.623	0.7293	17
k_{S2} g.dm^{-3}	1.2589	1.1803	1.0762	1.1718	8
$\alpha_1/Y_{X2/S1}$, [−]	3.871	3.3262	4.3868	3.8613	14
$\beta 1$ h^{-1}	0.0084	0.0084	0.0096	0.0088	8
$\alpha_1/Y_{X2/S2}$, [−]	0.0083	0.0153	0.01406	0.0125	30
β_2 h^{-1}	0.0166	0.0121	0.0098	0.0128	27
γ h^{-1}	0.0074	0.0075	0.0077	0.0076	2

Initial concentrations are in g dm^{-3}
Experiment 1: $X_2^\circ = 0.105$; $S_1^\circ = 10.84$; $S_2^\circ = 5.42$; $P_2^\circ = 0.35$. Experiment 2: $X_2^\circ = 0.101$; $S_1^\circ = 13.69$; $S_2^\circ = 4.84$; $P_2^\circ = 0.425$. Experiment 3: $X_2^\circ = 0.099$; $S_1^\circ = 19.85$; $S_2^\circ = 4.85$; $P_2^\circ = 0.38$. $Y_{P/S1} = Y_{P/S2} = 0.511$ (g/g)

a pretty good agreement between the experimental and model data, particularly for ethanol evolution and decay, due to formation of further products, e.g., acetaldehyde and acetic acid.

5 Conclusions

The presented mathematical modeling does not claim for generality. It does not include the use of more complicated substrates, like oligosaccharides and their conversion to monosaccharides. However, it can be extended and applied for various cases of substrates and microbial strains. What is important, the model considers the

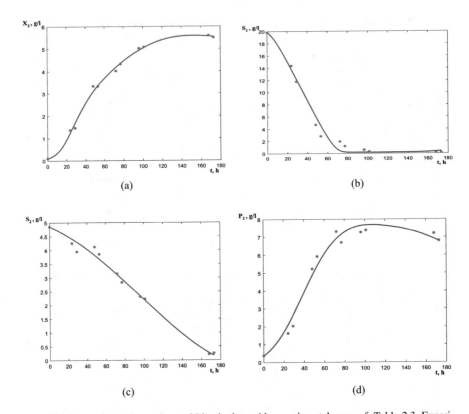

Fig. 2.5 Comparison of experimental kinetic data with experimental ones, cf. Table 2.3. Experiment 3. Biomass (*P. stipitis*), substrate, and product time profiles for different dilution rates. (a) biomass; (b) glucose; (c) xylose; (d) ethanol. Dr. $= 0.001$ h^{-1}

ethanol like intermediate and can predict its further conversion to other products, e.g., acetaldehyde and acetic acid. That is why one of the main merits of the model is its ability to predict the maximum ethanol concentration and the moment when the fermentation must be stopped to avoid losses of product. The model is applicable for multiple substrates, like glucose and xylose involving different microbial strains for each of the substrates.

The model is applicable either for batch or continuous processes taking place in two consecutively arranged reactors for separation of the processes with two different substrates and different strains.

Acknowledgement This work was supported by grant DFNI E02/16 of the Fund for Scientific research, Republic of Bulgaria.

References

1. B. Joshi, M.R. Bhatt, D.A. Sharma, J. Joshi, R. Malla, L. Sreerama, Lignocellulosic ethanol production: Current practices and recent developments. Biotechnol. Mol. Biol. Rev. **6**(8), 172–182 (2011)
2. B. Yang, C.E. Wyman, Pretreatment: The key to unlocking low-cost cellulosic ethanol. Biofuels Bioprod. Biorefin. **2**, 26–40 (2008)
3. S.T. Moe, K.K. Janga, T. Hertzberg, M.-B. Hägg, K. Øyaas, N. Dyrset, Saccharification of lignocellulosic biomass for biofuel and biorefinery applications a renaissance for the concentrated acid hydrolysis? Energy Procedia **20**, 50–58 (2012)
4. B.C. Saha, M.A. Cotta, Ethanol production from alkaline peroxide pretreated enzymatically saccharified wheat straw. Biotechnol. Prog. **22**, 449–453 (2006)
5. A. Kroumov, M. Zaharieva, V. Beshkov, Ethanol from cellulosic biomass with emphasis of wheat straw utilization. Analysis of strategies for process development. Int. J. Bioautom. **19**(4), 483–506 (2015)
6. K.V. Sarkanen, Lignin, in *Handbook of Pulp and Paper Technology*, ed. by K. W. Britt, (Van Nostrand Reinhold, New York, 1970)
7. N. Leksawasdi, E.L. Joachimsthal, P.L. Rogers, Mathematical modelling of ethanol production from glucose/xylose mixtures by recombinant Zymomonas mobilis. Biotechnol. Lett. **23**, 1087–1093 (2001)
8. J.P.A. Silva, S.I. Mussatto, I.C. Roberto, J.A. Teixeira, Ethanol production from xylose by Pichia stipitis NRRl Y-7124 in a stirred tank bioreactor. Braz. J. Chem. Eng. **28**(01), 151–156 (2011)
9. J.D. McMillan, *Xylose Fermentation to Ethanol: A Review* (National Renewable Energy Laboratory, National Technical Information Service. U.S. Department of Commerce, New York, 1993), pp. 30–36
10. M. Balat, Production of bioethanol from lignocellulosic materials via the biochemical pathway: A review. Energy Convers. Manag. **52**, 858–875 (2011)
11. S.C. Oliveira, R.C. Oliveira, M.V. Tacin, E.A.L. Gattás, Kinetic Modeling and optimization of a batch ethanol fermentation process. Bioprocess. Biotech. **6** (2016). https://doi.org/10.4172/2155-9821.1000266
12. C. Rossi, M. Porcelli, C. Mocenni, N. Marchettini, S. Loiselle, S. Bastianoni, A modelling approach for the analysis of xylose–ethanol bioconversion. Ecol. Model. **113**, 157–162 (1998)
13. S. Krahulec, B. Petschacher, M. Wallner, K. Longus, M. Klimacek, B. Nidetzky, Fermentation of mixed glucose-xylose substrates by engineered strains of Saccharomyces cerevisiae: Role of the coenzyme specificity of xylose reductase, and effect of glucose on xylose utilization. Microb. Cell Factories **9**, 16 (2010). https://doi.org/10.1186/1475-2859-9-16
14. K. Okamoto, A. Uchii, R. Kanawaku, H. Yanase, *Bioconversion of Xylose, Hexoses and Biomass to Ethanol by a New Isolate of the White Rot Basidiomycete Trametes Versicolor*, Springerplus (Cham, 2014)
15. S. Sreemahadevan, V. Singh, P. Kumar, R. Shaikh, Z. Ahammad, Mathematical modeling, simulation and validation for co-fermentation of glucose and xylose by Saccharomyces cerevisiae and Scheffersomyces stipites. Biomass Bioenergy **110**, 17–24 (2018) https://www.sciencedirect.com/science/article/pii/S096195341830014X

Chapter 3
Modeling and Simulation of Chemical Processes in Industrial Column Apparatuses

B. Boyadjiev and Christo Boyadjiev

Abstract A new approach for the chemical processes modeling in column apparatuses is presented in industrial column apparatuses. An exact approach for solutions of the equations in the convective type models is used. The use of experimental data, for the average concentration at the column end, for a concrete process and column, permits to be obtained the model parameters, related with the radial non-uniformity of the velocity. These parameter values permit to be used the average-concentration model for modeling of chemical processes with different reaction rates.

Key words Convection-diffusion model · Velocity components · Average-concentration model · Parameters identification

1 Introduction

The modeling and simulation of the chemical processes in column apparatuses is possible, using the new approach [1–3] on the basis of the physical approximations of the mechanics of continua, where the mathematical point (in the phase volume or on the surface between the phases) is equivalent to a small (elementary) physical volume, which is sufficiently small with respect to the apparatus volume, but at the same time sufficiently large with respect to the intermolecular volumes in the medium. On this base are presented convection-diffusion and average-concentration type models.

The convection-diffusion type models permit the qualitative analysis of the processes. These models are the base of the average-concentration models, which allow a quantitative analysis of the chemical processes in column apparatuses [4].

In the case of the chemical reactions in the industrial column apparatuses, the effects of the radial component of velocity and the axial diffusion transfer are not negligible and must be taken into account in the convection-diffusion and

B. Boyadjiev (✉) · C. Boyadjiev
Institute of Chemical Engineering, Bulgarian Academy of Sciences, Sofia, Bulgaria

© The Author(s), under exclusive license to Springer Nature Switzerland AG 2022
C. Boyadjiev (ed.), *Modeling and Simulation in Chemical Engineering*, Heat and
Mass Transfer, https://doi.org/10.1007/978-3-030-87660-9_3

average-concentration type models [6]. The using of the perturbations method [4, 5] permits to be obtained approximations solutions of the model equations. In this paper will be presented an exact approach for solutions of the equations in the convective-type models of industrial chemical processes in column apparatuses.

2 Convection-Diffusion Model

A theoretical analysis of the effect of the radial velocity components and the axial diffusion transfer in the industrial column chemical reactors will be presented in the case, when the radial velocity component is not equal to zero for pseudo-first order chemical reactions. In the stationary case, the convection-diffusion model [3, 4] has the form:

$$u\frac{\partial c}{\partial z} + v\frac{\partial c}{\partial r} = D\left(\frac{\partial^2 c}{\partial z^2} + \frac{1}{r}\frac{\partial c}{\partial r} + \frac{\partial^2 c}{\partial r^2}\right) - kc;$$

$$r = 0, \quad \frac{\partial c}{\partial r} \equiv 0; \quad r = r_0, \quad \frac{\partial c}{\partial r} \equiv 0; \quad z = 0, \quad c \equiv c^0, \quad u^0 c^0 \equiv uc^0 - D\frac{\partial c}{\partial z}.$$

$$(3.1)$$

$$\frac{\partial u}{\partial z} + \frac{\partial v}{\partial r} + \frac{v}{r} = 0; \quad r = r_0, \quad v(r_0, z) \equiv 0, \quad z = 0, \quad u = u(r, 0). \qquad (3.2)$$

In (3.1, 3.2) $c(r, z)$, D, are the concentrations [kg-mol·m^{-3}] and the diffusivities [m^2·s^{-1}] of the reagents in the fluid, $u(r, z)$ and $v(r, z)$—the axial and radial velocity components [m·s^{-1}], (r, z)—the radial and axial coordinates [m], k—chemical reaction rate constant, u^0, c^0—input ($z = 0$) velocity and concentrations.

The theoretical analysis of the model (3.1, 3.2) will be made, using generalized variables [1]:

$$r = r_0 R, \quad z = lZ, \quad \varepsilon = \frac{r_0}{l}, \quad c(r, z) = c(r_0 R, lZ) = c^0 C(R, Z),$$

$$u(r, z) = u(r_0 R, lZ) = u^0 U(R, Z), \quad v(r, z) = v(r_0 R, lZ) = u^0 \varepsilon V(R, Z).$$

$$(3.3)$$

As a result, from (3.1–3.3) is possible to be obtained:

$$U\frac{\partial C}{\partial Z} + V\frac{\partial C}{\partial R} = \text{Fo}\left(\varepsilon^2 \frac{\partial^2 C}{\partial Z^2} + \frac{1}{R}\frac{\partial C}{\partial R} + \frac{\partial^2 C}{\partial R^2}\right) - \text{Da}C;$$

$$R = 0, \quad \frac{\partial C}{\partial R} \equiv 0; \quad R = 1, \quad \frac{\partial C}{\partial R} \equiv 0; \quad Z = 0, \quad C \equiv 1, \quad 1 \equiv U - \varepsilon^2 \text{Fo}\frac{\partial C}{\partial Z}.$$

$$(3.4)$$

$$\frac{\partial U}{\partial Z} + \frac{\partial V}{\partial R} + \frac{V}{R} = 0; \quad R = 1, \quad V(1, Z) \equiv 0; \quad Z = 0, \quad U = U(R, 0). \quad (3.5)$$

In (3.5) are used the parameters:

$$Fo = \frac{D_i l}{u^0 r_0^2}, \quad Da = \frac{kl}{u^0}, \quad (3.6)$$

where Fo and Da are the Fourier and Damkohler numbers, respectively. In industrial conditions, the parameters $Fo < 10^{-2}$ are small and the model (3.4) has a convective form:

$$U\frac{\partial C}{\partial Z} + V\frac{\partial C}{\partial R} = -DaC;$$

$$Z = 0, \quad C \equiv 1; \quad R = 1, \quad \frac{\partial C}{\partial R} \equiv 0. \quad (3.7)$$

3 Axial and Radial Velocity Components

The theoretical analysis of the change in the radial non-uniformity of the axial velocity component (effect of the radial velocity component) in a column can be made by an appropriate hydrodynamic model, where the average velocity at the cross section of the column is a constant (inlet average axial velocity component), while the radial non-uniformity of the axial velocity component decreases along the column height and as a result a radial velocity component is initiated. In generalized variables (3.3) is possible to be used the model:

$$U = (2 - 0.4Z) - 2(1 - 0.4Z)R^2, \quad V = 0.2(R - R^3), \quad (3.8)$$

where the velocity components satisfy the Eq. (3.5). The velocity components (3.8) are presented on Figs. 3.1 and 3.2.

Eq. (3.7) is solved using the Method of Lines [6], by discretizing it with respect to $R_i = i/n, \quad i = 0, \ldots, n$, with 3 points central finite difference scheme, thus transforming it from partial differential equation to a system of ordinary differential equations with respect to Z:

$$U(R_0, Z)\frac{dC_0}{dZ} = -DaC_0;$$

$$U(R_i, Z)\frac{dC_i}{dZ} = -V(R_i)\frac{C_{i+1} - C_{i-1}}{R_{i+1} - R_{i-1}} - DaC_i;$$

$$U(R_n, Z)\frac{dC_n}{dZ} = -DaC_n;$$

$$Z = 0, \quad C_0 \equiv 1, \quad C_i \equiv 1, \quad C_n \equiv 1; \quad i = 1, \ldots, n - 1. \quad (3.9)$$

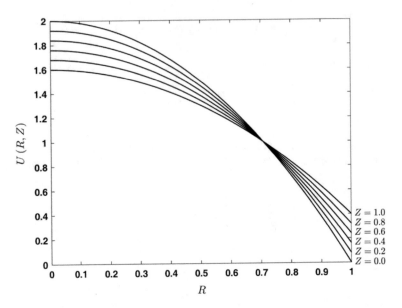

Fig. 3.1 Axial velocity component $U(R, Z)$ for different $Z = 0,\ 0.2,\ 0.4,\ 0.6,\ 0.8,\ 1.0$

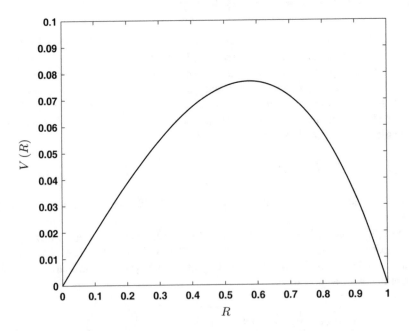

Fig. 3.2 Radial velocity component $V(R)$

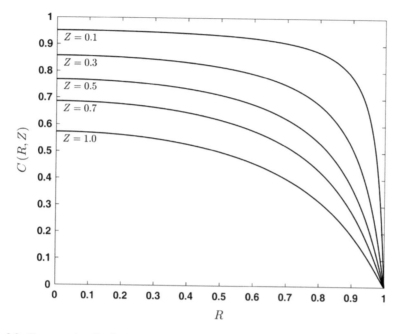

Fig. 3.3 Concentration distributions $C(R, Z)$ for different $Z = 0.1,\ 0.3,\ 0.5,\ 0.7,\ 1.0$

The problem (3.9) is stiff, that's why it is solved using MATLAB variable-step, variable-order solver ode15s, using second order backward differentiation formulas (BDF, also known as Gear's method).

The solutions of (3.7) in the case Da $= 1$ is presented on Fig. 3.3.

4   Average-Concentration Model

The functions $u(r, z),\ v(r, z),\ c(r, z)$ in (3.1) can be presented with the help of the average values of the velocity and concentration at the column cross-sectional area [1–3]:

$$\bar{u} = \frac{2}{r_0^2} \int_0^{r_0} r u(r)dr, \quad \bar{c}(z) = \frac{2}{r_0^2} \int_0^{r_0} r c(r, z)dr, \tag{3.10}$$

i.e.,

$$u(r, z) = \bar{u}\, U(R, Z), \quad v(r, z) = \varepsilon \bar{u} V(R), \quad c(r, z) = \bar{c}(z)\, \tilde{c}(r, z). \tag{3.11}$$

As a result, the following is obtained [3]:

$$\alpha(z)\bar{u}\frac{d\bar{c}}{dz} + [\beta(z) + \varepsilon\gamma(z)]\bar{u}\bar{c}_i = D\frac{d^2\bar{c}}{dz^2} - k\bar{c};$$

(3.12)

$$z = 0, \quad \bar{c} \equiv c^0, \quad \frac{d\bar{c}}{dz} \equiv 0,$$

where

$$\alpha(z) = \frac{2}{r_0^2}\int_0^{r_0} rU\tilde{c}\,dr, \quad \beta(z) = \frac{2}{r_0^2}\int_0^{r_0} rU\frac{\partial\tilde{c}}{\partial z}\,dr, \quad \gamma(z) = \frac{2}{r_0^2}\int_0^{r_0} rV\frac{\partial\tilde{c}}{\partial r}\,dr,$$

(3.13)

$$\tilde{c}(r,z) = \tilde{C}(R,Z), \quad U = U(R,Z), \quad V = V(R).$$

The theoretical analysis of the model (3.12) will be made, using the next generalized variables and functions:

$$z = lZ, \quad r = r_0R, \quad \bar{c}(z) = c^0\overline{C}(Z), \quad \overline{C}(Z) = 2\int_0^1 RC(R,Z)\,dR,$$

$$\tilde{c}(r,z) = \frac{c(r,z)}{\bar{c}(z)} = \frac{C(R,Z)}{\overline{C}(Z)} = \tilde{C}(R,Z), \quad \alpha(z) = A(Z) = 2\int_0^1 RU(R,Z)\tilde{C}(R,Z)\,dR,$$

$$\beta(z) = \beta(lZ) = B(Z) = 2\int_0^1 RU(R,Z)\frac{\partial\tilde{C}}{\partial Z}\,dR, \quad \gamma(z) = \gamma(lZ) = G(Z) = 2\int_0^1 RV(R)\frac{\partial\tilde{C}}{\partial R}$$

(3.14)

and as a result:

$$A(Z)\frac{d\overline{C}}{dZ} + [B(Z) + G(Z)]\overline{C} = \varepsilon^2\mathrm{Fo}\frac{d^2\overline{C}}{dZ^2} - \mathrm{Da}\overline{C};$$

(3.15)

$$Z = 0, \quad \overline{C} = 1, \quad \frac{d\overline{C}}{dZ} = 0.$$

In industrial conditions $\mathrm{Fo} < 10^{-2}$ and the model (3.15) has the convective form:

$$A(Z)\frac{d\overline{C}}{dZ} + [B(Z) + G(Z)]\overline{C} = -\mathrm{Da}\overline{C};$$

(3.16)

$$Z = 0, \quad \overline{C} = 1.$$

The solution of (3.7) and (3.14) permits to be obtained the average concentrations ("theoretical" values) $\overline{C}(Z_n)$ and functions $A(Z_n)$, $B(Z_n)$, $G(Z_n)$, $Z_n = 0.1$ $(n + 1)$, $n = 0, 1, \ldots, 9$, which are presented (points) on Figs. 3.4 and 3.5.

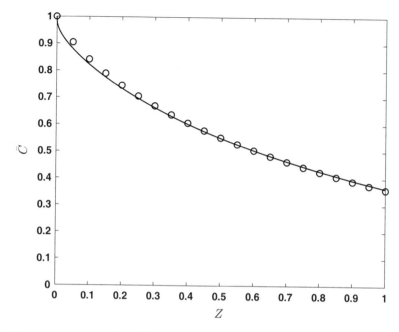

Fig. 3.4 Average concentrations $\overline{C}(Z)$: "theoretical" values $\overline{C}(Z_n)$, $Z_n = 0.1(n+1)$, $n = 0$, 1, ..., 9 (points); solution of (3.18) (lines)

From Fig. 3.5 is seen, that the functions $A(Z)$, $B(Z)$, $G(Z)$ are possible to be presented as the next approximations:

$$A(Z) = 1 + a_1 Z + a_2 Z^2, \quad B(Z) = b_1 Z^{b_2}, \quad G(Z) = gZ. \qquad (3.17)$$

As a result, the model (3.16) has the form:

$$\left(1 + a_1 Z + a_2 Z^2\right)\frac{d\overline{C}}{dZ} + \left(b_1 Z^{b_2} + gZ\right)\overline{C} = -\mathrm{Da}\overline{C};$$
$$Z = 0, \quad \overline{C} = 1, \qquad (3.18)$$

where the parameters $P(a_1, a_2, b_1, b_2, g)$ must be obtained using experimental data.

5 Parameters Identification

The value of the function $\overline{C}(1)$ obtained from (3.7) and (3.14) permits to be obtained the artificial experimental data $\overline{C}_{\exp}^m(1)$ for the column end ($Z = 1$):

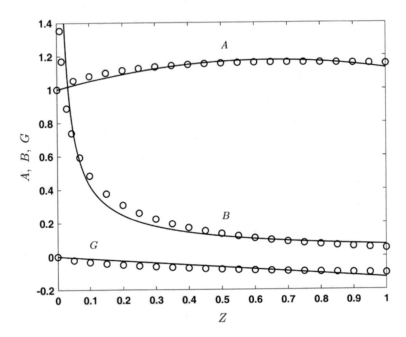

Fig. 3.5 Functions $A(Z_n)$, $B(Z_n)$, $G(Z_n)$, $Z_n = 0.1(n+1)$, $n = 0, 1, \ldots, 9$ (points) and theirs quadratic and linear approximations (3.17) (lines)

$$\overline{C}_{\exp}^m(1) = (0.95 + 0.1 B_m)\overline{C}(1), \quad m = 1, \ldots, 10, \tag{3.19}$$

where $0 \leq B_m \leq 1$, $m = 0, 1, \ldots, 10$ are obtained by a generator of random numbers.

The obtained artificial experimental data (3.19) is possible to be used for the illustration of the parameters P identification in the average-concentration model (3.18) by the minimization of the least-squares function:

$$Q(P) = \sum_{m=1}^{10} \left[\overline{C}(1, P) - \overline{C}_{\exp}^m(1) \right]^2, \tag{3.20}$$

where the values of $\overline{C}(1, P)$ are obtained after the solution of (3.17) for $Z = 1$.

The obtained ("experimental") parameter values (Table 3.1) are used for the solution of (3.18) and the results (the lines) are compared with the average ("theoretical") concentration values $\overline{C}(Z_n)$, $Z_n = 0.1(n+1)$, $n = 0, 1, \ldots, 9$ (points) on Fig. 3.4.

Table 3.1 Parameters $P(a_1, a_2, b_1, b_2, g)$

Parameters	"Theoretical" values	"Experimental" values
a_1	0.5373	0.5850
a_2	−0.4118	−0.4226
b_1	0.0695	0.0697
b_2	−0.7878	−0.6726
g	−0.1274	−0.1361

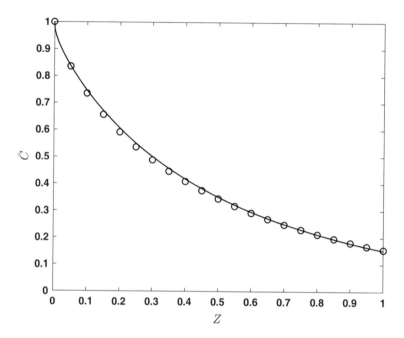

Fig. 3.6 Average concentrations $\overline{C}(Z)$ for Da $= 2$: "theoretical" values $\overline{C}(Z_n)$, $Z_n = 0.1(n + 1)$, $n = 0, 1, \ldots, 9$ (points); solution of (3.18) (lines)

6   Effect of the Chemical Reaction Rate

The effect of the chemical reaction rate will be obtained in the case, when Da $= 2$ in the models (3.7) and (3.18). The solution of (3.7) and (3.14) permits to be obtained theoretical values of the average concentration (points), which are compared (Fig. 3.6) with the solution of the average-concentration model Eq. (3.18), where the obtained experimental parameters values (Table 3.1) are used.

7 Conclusions

A new approach for the chemical processes modeling in column apparatuses is presented in industrial column apparatuses. An exact approach for solutions of the equations in the convective type models is used. The use of experimental data, for the average concentration at the column end, for a concrete process and column, permits to be obtained the model parameters, related with the radial non-uniformity of the velocity. These parameter values permit to be used the average-concentration model for modeling of chemical processes with different reaction rates.

Acknowledgement This work is supported by Project of Fundamental Scientific Research 19-58-18004, conducted by RFBR and the National Science Fund of Bulgaria under contract No KP 06 RUSIA-3 from 27 Sep. 2019, "Modeling, simulation and experimental investigations of the interphase mass transfer and separation in distillation, absorption, adsorption and catalytic processes in industrial column apparatuses" and contract No KP 06-H37/11/ 06.12.2019 "Integrated absorption-adsorption process for waste free decontamination of gases from sulfur dioxide."

References

1. C. Boyadjiev, *Theoretical Chemical Engineering. Modeling and Simulation* (Springer, Berlin, 2010)
2. C. Boyadjiev, M. Doichinova, B. Boyadjiev, P. Popova-Krumova, *Modeling of Column Apparatus Processes* (Springer, Berlin, 2016)
3. C. Boyadjiev, M. Doichinova, B. Boyadjiev, P. Popova-Krumova, *Modeling of Column Apparatus Processes*, 2nd edn. (Springer, Berlin, 2018)
4. B. Boyadjiev, C. Boyadjiev, New models of industrial column chemical reactors. Bulg. Chem. Commun. **49**(3), 706–710 (2017)
5. B. Boyadjiev, C. Boyadjiev, New approach to modelling and simulation of chemical and mass transfer processes in column apparatuses. Bulg. Chem. Commun. **50**, 7–23 (2018)
6. W.E. Schiesser, *The Numerical Method of Lines* (Academic Press, San Diego, 1991)

Chapter 4
Multi-Period Deterministic Model of Sustainable Integrated of Hybrid First and Second Generation Bioethanol Supply Chains for Synthesis and Renovation

Boyan B. Ivanov and Yunzile R. Dzhelil

Abstract This paper focuses on designing mathematical model of an integrated bioethanol supply chain (IBSC) that will account for economic, environmental, and social aspects of sustainability. A mixed integer linear programming (MILP) model is proposed to design an optimal IBSC. Bioethanol production from renewable biomass has experienced increased interest in order to reduce Bulgarian dependence on imported oil and reduce carbon emissions. Concerns regarding cost efficiency and environmental problems result in significant challenges that hinder the increased bioethanol production from renewable biomass. The model considers key supply chain activities including biomass harvesting/processing and transportation. The model uses the delivered feedstock cost, energy consumption, and GHG emissions as system performance criteria. The utility of the supply chain simulation model is demonstrated by considering a biomass supply chain for a biofuel facility in Bulgarian scale. The results show that the model is a useful tool for supply chain management, including selection of the optimal bioethanol facility location, logistics design, inventory management, and information exchange.

Key words Bioethanol supply chain · Mathematical model · Economic · Environmental and social aspects

1 Introduction

Biofuel production and use is promoted worldwide. Its use could potentially reduce emissions of greenhouse gases and the need for fossil fuels [1]. Accordingly, the European Union has imposed a mandatory target of 10% biofuel production by the year 2020 [2]. Biofuels are produced from biomass feedstocks. Their use for energy purposes has the potential to provide important benefits. Burning biofuel releases as

B. B. Ivanov · Y. R. Dzhelil (✉)
Institute of Chemical Engineering, Bulgarian Academy of Sciences, Sofia, Bulgaria

much CO_2 as the amount that has been absorbed by the biomass in its formation. Another advantage of biomass is its availability in the world due to its variety of sources. Despite its advantages, increasing quantities of biofuels to achieve EC objectives are accompanied by growing quantities of waste products. These wastes are related to the biofuel's lifecycle from crop cultivation, transportation, and production up to distribution and use. The main liquid biofuels are bioethanol and biodiesel. Depending on the raw material used, production is considered in two generations.

The first generation used as feedstock crops containing sugar and starch to produce bioethanol [3]. In the production of bioethanol, the advantage of these materials is that they can be grown on contaminated and saline soils, as the process does not affect the fuel production. The drawback is that they raise issues related to their competitiveness in the food sector. Excessive use of fertilizers, pesticides, and chemicals to grow them also leads to accumulation of pollutants in groundwater that can penetrate into water courses and thus degrade water quality.

Referring to the second generation, bioethanol is produced by using as raw material waste biomass (agricultural and forest waste) [4], i.e., lignocellulose which is transformed into a valuable resource as bioethanol.

The main technologies for production of bioethanol are fermentation, distillation, and dehydration [5]. The wastes of biofuels are divided into production and performance. The technological waste is produced mainly in generation of products that occur as waste. The management of such waste is related to their reduction, recovery, and disposal.

The present study deals with the issue of designing an optimal Integrated Bioethanol Supply Chains (IBSC) model for waste management in the process of biofuel production and use. Tools have been developed for the formulation of a mathematical model for the description of the parameter, the restrictions, and the goal function.

2   Literature Review

Literature review [6] shows that a significant amount of research is being conducted to design a bioethanol supply chain. Following are the up-to-date literature on **first generation** bioethanol supply chain, **second generation** bioethanol supply chain, hybrid generation bioethanol supply chain, industrial symbiosis, and sustainability respectively.

First generation bioethanol is produced from food-based biomass, such as corn and soybean [7]. In [8] a deterministic model to design a cost-effective corn-based bioethanol supply chain is developed. Later, the authors [9] added environmental criterion along with the economic one to this model. The objective is to simultaneously improve economic and environmental aspects of sustainability. The study indicates that the strategic, tactical, and operational decisions are highly sensitive to the environmental considerations; and the **first generation** bioethanol production can put bioethanol supply chain into risk under proposed European Union (EU) environmental standards.

Second generation bioethanol is produced from lignocellulosic-based biomass, such as woody materials, crop residuals, or dedicated energy crops (e.g. switchgrass, miscanthus, etc.). In [10] a deterministic model to design a cost-effective lignocellulosic bioethanol supply chain is developed. A deterministic model to design a cost-effective **second generation** bioethanol supply chain is proposed in [11], which aims to determine optimal number, size, and location of bioethanol plants. In [12] a MILP model to design lignocellulosic bioethanol supply chain is developed. Also, in [13] a deterministic model to design a lignocellulosic bioethanol supply chain in order to maximize the profit of bioethanol supply chain is considered.

In [14] a deterministic model to design the optimal switchgrass-based supply chain to minimize the total cost is developed. The authors of [15] proposed a deterministic model to design an economically sustainable **second generation** bioethanol supply chain under imperfect market competition.

Hybrid generation bioethanol supply chain (HGBSC) is produced from a combination of **first generation** bioethanol and **second generation** biomass. In [16] a deterministic MILP model to design an economically and environmentally sustainable HGBSC is developed. Authors of [17] developed a deterministic model to design an HGBSC. The objective is to improve economic aspect under environmental restrictions. While, later they [18] proposed a deterministic model to design a HGBSC that aims to improve economic benefits under food crop and land use restrictions. In [19] a deterministic model to design HGBSC is developed. The research aims to improve economic and environmental aspects of sustainability. The results suggest that strong environmental restrictions highly impact the technological choices and bioethanol supply chain design. In [20] deterministic model to design a cost-effective HGBSC is represented. The proposed model aims to design new second generation bioethanol supply chain while considering the existing **first generation** bioethanol supply chain. In [21] a deterministic model to design HGBSC that aims to improve economic benefits under emission trading schemes (i.e., environmental aspect of sustainability) is developed. In [22] a multi-objective MILP model to design HGBSC, which improves economic and environmental (carbon and water footprint) performance of HGBSC, is considered.

2.1 Summary of Literature Review

While considerable amount of research has been conducted in the area of bioethanol supply chain and IS, there are several important issues missing in the existing literature. These include:

1. None of the up-to-date literature has focused on designing HGBSC that aims to simultaneously improve several sustainability indicators such as profit, greenhouse gas emissions, irrigation land usage, water usage, and energy efficiency, under uncertainties such as demand, price, and biomass yield.
2. None of the up-to-date literature has studied the impact of sustainability standards or regulations on the design of HGBSC.
3. None of the up-to-date literature has considered existing bioethanol plant configurations and their capacities while integrating or designing new bioethanol.
4. None of the update literature has considered exploiting the sustainability benefits of IS while designing a bioethanol supply chain.

Therefore, in order to bridge the gap in the literature, this paper focuses on designing sustainable integrated hybrid generation bioethanol supply chains in which the sustainability benefits of sustainability indicators are exploited while designing sustainable integrated hybrid generation bioethanol supply chains. In addition, existing bioethanol plant configurations and their capacities are considered in the study to meet the future bioethanol requirements.

3   Objectives of the Present Study

The main focus of the strategic level is the design of an efficient supply chain to minimizing total costs of the process and to provide the best scenario for reducing the environmental impact of the whole supply chain. This decision level prescribes the source of the raw material to ensure an effective and efficient configuration of the supply chain, the suitable technology used in the process, the location and capacity of the processing plant, and the sustainability issues. The present study deals with the issue of designing optimal integrated supply chains (SCs) for waste management in the process of biofuel production and usage. Tools have been developed for formulation of a mathematical model for description of the parameter, the restrictions and the goal function.

4   Problem Definition

This research studies a proposed echelons and possible links of IBSC as shown in Fig. 4.1. A list of indices, parameters, and decision variables is given in the **Appendix** Nomenclature section at the end of the article. In this work an IBSC involving raw material sites, production plants, and customer zones is considered.

In short, the SC and plant design problem consists in determining simultaneously the following logistics decisions need to be optimized:

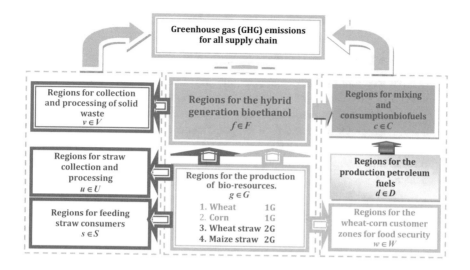

Fig. 4.1 Superstructure integrated bioethanol supply chain (IBSC)

- cultivation sites selection from the i biomass supply zones and allocation of available land for biomass production,
- amount of crop residues to be purchased from the i biomass supply zones,
- amount of biomass to be directly sold from the biomass supply zones,
- material flow of lignocellulosic feedstock from the biomass supply zones to the biorefineries,
- sites selection from biorefinery locations,
- volume of bioethanol to be produced by the biorefineries,
- volume of bioethanol to be directly sold from the biorefineries,
- material flow of bioethanol from the biorefineries to the biofuel demand zones.

The problem addressed in this work can be formally stated as follows. Given are a set of biofuel crops that can be converted to biofuel. These include agricultural feedstocks, e.g. wheat, corn, energy crops, etc. A planning horizon of 1 year for government regulations including manufacturing, construction, and carbon tax is considered. A Biofuel Supply Chain (BSC) network superstructure including a set of harvesting sites and a set of demand zones, as well as the potential locations of a number of collection facilities and bio refineries are set. Data for biofuel crops production and harvesting are also given. For each demand zone, the biofuel demand is given, and the environmental burden associated with biofuel distribution in the local region is known. For each transportation link, the transportation capacity, available transportation modes, distance, and emissions of each transportation type are known.

4.1 General Formulation of the Problem

The overall problem can be summarized, as follows:

- Potential locations of fuel demand centers and their biofuel demand,
- Demand for petroleum fuel for each of the demand centers for fuel,
- The minimum required ratio between petroleum fuel and biofuel for blending,
- Biomass feedstock types and their geographical availability,
- Specific Greenhouse Gas (GHG) emission factors of the biofuel life cycle stages,
- Potential areas where systems for utilization of solid waste from production can be installed.

The objectives are to optimize the economic, environmental, and social performances of Integrated Bioethanol Supply Chains by determination the following decision variables:

- Supply chain network structure,
- Locations and scales of biofuel production facilities and biomass cultivation sites,
- Flows of each biomass type and biofuel between regions,
- Modes of transport for delivery for biomass and biofuel (B100),
- The GHG emissions for each stage in the life cycle,
- Supply strategy for biomass to be delivered to production facilities,
- Distribution processes for biofuel to be sent to demand zones.

5 Model Formulation

The role of the optimization model is to identify what combination of options is the most efficient approach to supply the facility. The problem for the optimal location of bioethanol (B100) production plants and the efficient use of the available land is formulated as a MILP model with the following notation:

5.1 Mathematical Model Description

To start with the description of the MILP model, we first introduce the parameters, that are constant and known a priori, and the variables that are subject to optimization. Then we describe step by step the mathematical model by presenting the objective function and all the constraints. First of all, we introduce the set of time intervals of the horizon of planning $t = \{1, 2, \ldots, T\}$. The subscript t indicates the variable or parameter corresponding to the tth interval of the planning.

In this part the mathematical model that is used in the network design is described. Before describing the mathematical model, the input parameters, the decision variables, and the sets, subsets, and indices are listed below **Appendix**.

5.2   Supply Chain Structure

The supply chain structure is the arrangement of different entities in the supply chain. It represents the connection between the suppliers and the final destination of the waste bioresources, represented in the form of material and mass flows. Among the types of supply chain structures, convergent and conjoined structures are the most studied. Figure 4.1 presents the supply chain structures that can be implemented in the case of the recycled waste bioresources.

Following, the Basic Relationships and Constraints are imposed and the definition of variables and parameters are presented below.

5.3   Basic Relationships

As noted above, the assessment of IBSC production and distribution of bioethanol (E100) will be made by economic, environmental, and social criteria.

5.3.1   Model of Environmental Assessment of IBSC

The environmental impact of the IBSC is measured in terms of total GHG emissions ($kg\ CO_2 - eq$) stemming from supply chain activities and the total emissions are converted to carbon credits by multiplying them with the carbon price in the market.

The environmental objective is to minimize the total annual GHG emission resulting from the operations of the IBSC. The formulation of this objective is based on the field-to-wheel life cycle analysis, which takes into account the following life cycle stages of biomass-based liquid transportation fuels:

- biomass cultivation, growth, and acquisition,
- biomass transportation from source locations to processing facilities,
- transportation of bioethanol (E100) facilities to the demand zones,
- local distribution of liquid transportation fuels in demand zones,
- emissions from bioethanol (E100) and gasoline usage in vehicle operations.

Ecological assessment criteria will represent the total environmental impact at work on IBSC through the resulting GHG emissions for each time interval $t \in T$. These emissions are equal to the sum of the impact that each of the stages of life cycle has on the environment. The GHG emission rate is generally defined as follows for each time interval $t \in T$:

$$TEI_t = ELS_t + ELB_t + ELD_t + ETT_t + ESW_t + ESTRAW_t + ECAR_t, \quad \forall t \quad (4.1)$$

The different components of the Eq. (4.1) are explained in the following.

TEI_t

Total GHG impact at work on IBSC [$kg\,CO_2 - eq\,d^{-1}$];

$$\left\{\begin{array}{l} ELS_t \\ ELB_t \\ ELD_t \\ ETT_t \end{array}\right\}$$

GHG impact of life cycle stages;

ESW_t

Emissions from utilization solid waste for each time interval $t \in T$;

$ESTRAW_t$

Emissions generated as a result of the utilization of residual straw in the regions $g \in G$ for each time interval $t \in T$;

$ECAR_t$

Emissions from bioethanol (E100) and gasoline usage in vehicle operations [$kg\,CO_2 - eq\,d^{-1}$];

Evaluation of environmental impact at every stage of life cycle is:

(a) *Growing biomass* ELS_t [$kg\,CO_2 - eq\,d^{-1}$];
(b) *Production of bioethanol l(E100)* ELB_t [$kg\,CO_2 - eq\,d^{-1}$];
(c) *Production of petroleum gasoline* ELD_t [$kg\,CO_2 - eq\,d^{-1}$];
(d) *Utilization of solid wastes* ESW_t [$kg\,CO_2 - eq\,d^{-1}$];
(e) *Transportation biomass for bioethanol production* ETA_t [$kg\,CO_2 - eq\,d^{-1}$];
(f) *Transportation bioethanol(E100)* ETE_t [$kg\,CO_2 - eq\,d^{-1}$];
(g) *Transportation gasoline* ETD_t [$kg\,CO_2 - eq\,d^{-1}$];
(h) *Transportation of solid waste* ETW_t [$kg\,CO_2 - eq\,d^{-1}$];
(i) *Transportation of straw and corn cobs* ETU_t [$kg\,CO_2 - eq\,d^{-1}$];
(j) *Transportation of wheat–corn for food security* ETV_t [$kg\,CO_2 - eq\,d^{-1}$];
(k) *Usage in vehicle operations bioethanol(E100) and gasoline* $ECAR_t$ [$kg\,CO_2 - eq\,d^{-1}$].

Greenhouse Gases to Grow Biomass ELS_t

GHG emissions resulting from the production of biomass depend on the cultivation practice adopted as well as on the geographical region in which the biomass crop has been established [9]. In particular, the actual environmental performance is affected by fertilizers and pesticides usage, irrigation techniques, and soil characteristics. The factor may differ strongly from one production region to another. Accordingly, the biomass production stage is defined as follows:

$$ELS_t = \sum_{i \in I} \sum_{g \in G} \left(EFBC_{igt} \frac{\beta_{igt} \left(A_{igt} + A^F_{igt} \right)}{\alpha_t} \right), \quad \forall \ t, \qquad (4.2)$$

where ELS_t denotes the total environmental impact of biomass cultivation, which in general represents the production rate of resource $i \in I$ in region $g \in G$, refers in this equation to the cultivation rate of biomass $i \in I$ in that region, $[kg\ CO_2 - eq\ d^{-1}]$.

Total GHG Emissions from Bioethanol (E100) Production ELB_t

The environmental impact of the bioethanol (E100) production stage is related to raw materials and the technology employed for the production of bioethanol (E100). Accordingly, the GHG emissions resulting from this life cycle stage were assumed proportional to the biomass-specific amount of biofuel produced:

$$ELB_t = \sum_{i \in I} \sum_{f \in F} \sum_{c \in C} \sum_{b \in B} \sum_{p \in P} \left(EFBP_{ip} QBP_{ifcbpt} + EFBPs_{ip} QBPs_{ifcbpt} \right), \quad \forall \ t \qquad (4.3)$$

where ELB_t is total environmental impact of bioethanol (E100) production.

Since only one of the technologies $p \in P$ can be selected for a region $f \in F$ (which is guaranteed by the condition $\sum_{p \in P} ZF_{pft} \leq 1.0 \ \forall \ t,f$), it QBP_{ifcbpt} is equal to "0" for all except $p \in P$ for the selected technology. This is ensured by implementing the inequality

$$G^{MAX} ZF_{pft} \geq QBP_{ifcbpt}, \quad \forall \ i, f, c, b, p, t$$

and

$$G^{MAX} ZF_{pft} \geq QBPs_{ifcbpt}, \quad \forall \ i, f, c, b, p, t$$

where G^{MAX} there is a large enough number.

Total GHG Emissions from Gasoline Production ELD_t

$$ELD_t = \sum_{d \in D} \sum_{c \in C} \sum_{b \in B} EDP_{dt} QD_{dcbt}, \quad \forall \ t \qquad (4.4)$$

where ELD_t is total environmental impact of gasoline production.

The Environmental Impact of Transportation ETT_t

The global warming impact related to both biomass supply and fuel distribution depends on the use of different transport means fuelled with fossil energy, typically either conventional oil-based fuels or electricity. The resulting GHG emissions of each transport option depend on both the distance run by the specific means and the freight load delivered. As a consequence, the emission factor represents the total carbon dioxide emissions equivalent accordingly:

$$ETT_t = ETA_t + ETB_t + ETD_t + ETW_t + ETU_t + ETV_t \qquad (4.5)$$

where ETT_t is environmental impact of transportation of resources;
$$ETA_t = \sum_{i\in I}\sum_{g\in G}\sum$$

$$\sum_{l\in L}\left(EFTMA_{il}ADG_{gfl}QI_{igflt}\right) + \sum_{i\in I}\sum_{g\in G}\sum_{f\in F}\sum_{e\in E}\left(EFTMAs_{ie}ADG_{gfl}QIs_{igfet}\right), \quad \forall\, t$$

is environmental impact of transportation biomass for bioethanol production;
$$ETB_t = \sum_{f\in F}\sum_{c\in C}\sum_{b\in B}\left(EFTB_bADF_{fcb}QB_{fcbt}\right), \quad \forall\, t \text{ is environmental impact of trans-}$$
portation bioethanol (E100) from zones $f \in F$ to $c \in C$ where $QB_{fcbt} =$

$$\sum_{i\in I}\sum_{p\in P}\left(QBP_{ifcbpt} + QBPs_{ifcbpt}\right);$$
$$ETD_t = \sum_{d\in D}\sum_{c\in C}\sum_{b\in B}\left(EFTB_bADD_{dcb}QD_{dcbt}\right), \quad \forall\, t \text{ is environmental impact of}$$
transportation gasoline from zones $d \in D$ to $c \in C$, $[kg\,CO_2 - eq\,d^{-1}]$;
$$ETW_t = \sum_{f\in F}\sum_{w\in W}\sum_{m\in M}\left(EFTRW_mADW_{fwm}QW_{fwmt}\right), \quad \forall\, t \text{ is environmental impact}$$
of transportation solid wastes from zones $f \in F$ to $w \in W$, $[kg\,CO_2 - eq\,d^{-1}]$;
$$ETU_t = \sum_{g\in Gu}\sum_{e\in E}\sum_{i\in I}\left(EFTRU_eADU_{gue}QU_{iguet}\right), \quad \forall\, t \text{ is environmental impact}$$
of transportation straw from zones $g \in G$ to $u \in U$, $[kg\,CO_2 - eq\,d^{-1}]$;
$$ETV_t = \sum_{g\in Gv}\sum_{z\in Z}\sum_{i\in I}\left(EFTRV_zADV_{gvz}QV_{igvzt}\right), \quad \forall\, t \text{ is environmental impact of}$$
transportation wheat–corn for food security from zones $g \in G$ to $v \in V$, $[kg\,CO_2 - eq\,d^{-1}]$.

Total GHG Emissions from Utilization Solid Wastes ESW_t, $[kg\,CO_2 - eq\,d^{-1}]$

$$ESW_t = \sum_{f\in F}\left(ESF1_{ft}\left(FSW_{ft} - FSWW_{ft}\right)\right) + ESWW_t, \quad \forall\, t \qquad (4.6)$$

where ESW_t are the GHG emissions that would be generated if part of the solid waste was not treated in the places designated for this purpose, $[kg\,CO_2 - eq\,d^{-1}]$ and

FSW_{ft} is the amount of solid waste generated during the operation of each of the plants $f \in F$ for the time interval $t \in T$ and $FSWW_{ft}$ is the amount of solid waste generated by $f \in F$ which is processed in all plants $w \in W$.

$$FSW_{ft} = \sum_{i \in I} \sum_{p \in P} \sum_{c \in C} \sum_{b \in B} \left(SW_{ip}QBP_{ifcbpt} + SWs_{ip}QBPs_{ifcbpt} \right), \quad \forall \ t, f$$

$$FSWW_{ft} = \sum_{m \in M} \sum_{w \in W} QW_{fwmt}, \quad \forall \ t, f$$

While $ESWW_t$ represents GHG emissions generated from solid waste disposal when it is carried out in one of the plants $w \in W$ and is determined according to the dependence:

$$ESWW_t = \sum_{f \in F} \sum_{w \in W} \sum_{m \in M} \sum_{s \in S} \left(ESW1_{wst}QWS_{fwmst} \right), \quad \forall \ t$$

Total GHG Emissions from Utilization Straw $ESTRAW_t$

$$ESTRAW_t = \sum_{i \in I} \left(ESU_i \left(\sum_{g \in G} \left(A_{igt} + A_{igt}^F \right) \frac{\beta s_{igt}}{\alpha_t} - \sum_{g \in G} \sum_{u \in U} \sum_{e \in E} QU_{iguet} - \sum_{g \in G} \sum_{f \in F} \sum_{e \in E} QIs_{igfet} \right) \right), \quad \forall \ t$$

$$(4.7)$$

where $ESTRAW_t$ are the GHG total emissions from utilization straw $[kg \ CO_2 - eq \ d^{-1}]$.

GHG Emissions from Bioethanol (E100) and Gasoline Usage in Vehicle Operations $ECAR_t$

$$ECAR_t = ECB \sum_{f \in F} \sum_{c \in C} \sum_{b \in B} QB_{fcbt} + ECG \sum_{d \in D} \sum_{c \in C} \sum_{b \in B} QD_{dcbt}, \quad \forall \ t, \qquad (4.8)$$

where $ECAR_t$ is GHG emissions from bioethanol (E100) and gasoline usage in vehicle operations.

5.3.2 Model of Economic Assessment of an IBSC TDC_t, [$ $year^{-1}$]

The annual operational cost includes the biomass feedstock acquisition cost, the local distribution cost of final fuel product, the production costs of final products, and the transportation costs of biomass, and final products. In the production cost, we consider both the fixed annual operating cost, which is given as a percentage of the corresponding total capital investment, and the net variable cost, which is proportional to the processing amount. In the transportation cost, both distance-fixed cost and distance-variable cost are considered. The economic criterion will be the cost of living expenses to include total investment cost of bioethanol (E100) production facilities and operation of the IBSC. This price is expressed through the dependence (4.8) for each time interval $t \in T$:

$$TDC_t = TIC_t + TIW_t + TPC_t + TPW_t + TTC_t + TTAXB_t - TL_t \\ - TS_t, \quad \forall t \tag{4.9}$$

where
   TDC_t
Total cost of an IBSC for year, [$ $year^{-1}$];
   TIC_t
Total investment costs of production capacity of IBSC relative to the operational period of redemption and up time of the plant per year, [$ $year^{-1}$];
   TIW_t
Total investment costs of solid waste production capacity of IBSC relative to the operational period of redemption and up time of the plant per year, [$ $year]^{-1}$;
   TPC_t
Production cost for biorefineries, [$ $year^{-1}$];
   TPW_t
Production cost for solid waste plants, [$ $year^{-1}$];
   TTC_t
Total transportation cost of an IBSC, [$ $year^{-1}$];
   $TTAXB_t$
A carbon tax levied according to the total amount of CO_2 generated in the work of IBDS, [$ $year^{-1}$];
   TL_t
Government incentives for bioethanol (E100) production and use, [$ $year^{-1}$];
   TS_t
Total costs of selling crop residues (straw and corn cobs) for other purposes, [$ $year^{-1}$].

Model Investment Costs for Biorefineries by Year TIC_t

A rational SC planning over the time is based upon the assumption that once a production facility has been built, it will be operating for the remaining time frame.

$$TIC_t = \varepsilon_t \sum_{f \in F} \sum_{p \in P} \left(Cost_{pf}^F Z_{pft} \right), \quad \forall \ t \tag{4.10}$$

where ε_t is a discount factor defined as in [23] and present by [24] ε_t is calculated by Eq. (4.11):

$$\varepsilon_t = \frac{1}{(1 + \varsigma_t)} \tag{4.11}$$

where ς_t is the interest rate [%] for time interval $t \in T$.

Capital cost of biorefinery for each region is determined by the equation:

$$Cost_{pf}^F = M_f^{\cos t} Cost_p, \quad \forall \ p \in P, \quad \forall \ f \in F, \tag{4.12}$$

where $M_f^{\cos t}$ is a correction factor in the price of biorefineries in the region $f \in F$ according to its installed $M_f^{\cos t} \geq 1$.

Model Investment Costs for Solid Waste Plants by Year TIW_t

The total cost of the utility plants to be constructed over a given time interval is determined according to the equation:

$$TIW_t = \varepsilon_t \sum_{w \in W} \sum_{s \in S} \left(Cost_{sw}^W ZW_{swt} \right), \quad \forall \ t \tag{4.13}$$

where

$$Cost_{sw}^W = M_w^M CostW_s, \quad \forall \ s \in S, \quad \forall \ w \in W, \tag{4.14}$$

And M_w^W is a correction factor in the price of solid waste plant in the region $w \in W$ according to its installed $M_w^W \geq 1$.

Total Production Cost Model of IBSC TPC_t

Total production cost term, TPC_t consists of biomass cultivation TPA_t, bioethanol production costs TPB_t, $TPBs_t$, and production cost for gasoline TPD_t as follows for each time interval $t \in T$:

$$TPC_t = TPA_t + TPB_t + TPBs_t + TPD_t, \quad \forall\ t \tag{4.15}$$

where the components of (4.15) are defined according to the relations:

$$\left.\begin{aligned}
TPA_t &= \sum_{i \in I}\sum_{g \in G}\left(UPC_{igt}\beta_{igt}\left(A_{igt} + A^F_{igt}\right)\right) \\
TPB_t &= \sum_{i \in I}\sum_{f \in Fc}\sum_{c \in Cb}\sum_{Bp \in P}\left(\alpha_t UPB_{ipft}QBP_{ifcbpt}\right) \\
TPBs_t &= \sum_{i \in I}\sum_{f \in Fc}\sum_{c \in Cb}\sum_{Bp \in P}\left(\alpha_t UPBs_{ipft}QBPs_{ifcbpt}\right) \\
TPD_t &= \sum_{c \in Cb}\sum_{d \in D}\left(\alpha_t UPD_{dt}QD_{dcbt}\right)
\end{aligned}\right\}, \quad \forall\ t$$

Total Utilization Cost Model of Solid Waste TPW_t

Total utilization cost of solid waste is calculated for equation:

$$TPW_t = \alpha_t \sum_{f \in F}\sum_{w \in W}\sum_{m \in M}\sum_{s \in S}\left(UPW_{swt}QWS_{fwmst}\right)$$
$$+ \sum_{f \in F}\left(QWPLANTS_{ft}UPSW_{ft}\right), \quad \forall\ t \tag{4.16}$$

Since only one of the size $s \in S$ can be selected for a region $w \in W$ (which is guaranteed by the condition $\sum_{s \in S}ZWF_{swt} \leq 1.0 \quad \forall\ t, w$), it QWS_{fwmst} is equal to "0" for all except $s \in S$ for the selected size. This is ensured by implementing the inequality $G^{MAX}ZWF_{swt} \geq QWS_{fwmst}, \quad \forall\ f, w, m, s, t$ where G^{MAX} there is a large enough number.

Where $QWFULL_{ft}$ is the total amount of solid waste [$ton\ year^{-1}$] generated by each biorefinery $f \in F$ and $QWPLANTS_{ft}$ is the total amount of solid waste that is treated in each of the plants $f \in F$.

$$\left.\begin{aligned}
QWPLANTS_{ft} &= QWFULL_{ft} - \alpha_t \sum_{s \in Sw}\sum_{w \in Wm}\sum_{e \in W} QWS_{fwmst} \\
QWFULL_{ft} &= \alpha_t \sum_{i \in I}\sum_{c \in Cb}\sum_{Bp \in P}\left(SW_{ipt}QBP_{ifcbpt} + SWs_{ipt}QBPs_{ifcbpt}\right)
\end{aligned}\right\}, \quad \forall\ t, f$$

Total Transportation Cost Model TTC_t

With regard to transports, both the biomass delivery to conversion plants and the fuel distribution and transport of gasoline to blending terminals are treated as an additional service provided by existing actors already operating within the industrial/transport infrastructure. As a consequence, TTC_t is evaluated as follows:

$$TTC_t = TTCA + TTCB_t + TTCD_t$$

$$+TTCW_t + TTCU_t + TTCV_t, \quad \forall \, t \qquad (4.17)$$

where
$TTCA$

$$= \sum_{l \in L} \sum_{i \in I} \sum_{f \in F} \sum_{g \in G} \left(\alpha_t UTC_{igfl} QI_{igflt} \right) + \sum_{e \in E} \sum_{i \in I} \sum_{f \in F} \sum_{g \in G} \left(\alpha_t UTCs_{igfe} QIs_{igfet} \right), \quad \forall \, t$$

is the transportation cost for energy crops and straw for bioethanol production,
$TTCB_t = \sum_{b \in Bc} \sum_{c \in C} \sum_{f \in F} \left(\alpha_t UTB_{fcb} QB_{fcbt} \right), \quad \forall \, t$ is the transportation cost for bioethanol (E100),
$TTCD_t = \sum_{b \in Bc} \sum_{c \in Cd} \sum_{c \in C} \left(\alpha_t UTD_{dcb} QD_{dcbt} \right), \quad \forall \, t$ is the transportation cost for gasoline,
$TTCW_t = \sum_{m \in Mw} \sum_{e \in W} \sum_{f \in F} \left(\alpha_t UTW_{fwm} QW_{fwmt} \right), \quad \forall \, t$ is the transportation cost for solid waste,
$TTCU_t = \sum_{e \in Eg} \sum_{e \in Gu} \sum_{e \in U} \sum_{i \in I} \left(\alpha_t UTU_{gue} QU_{iguet} \right), \quad \forall t$ is the transportation cost for straw collection and processing,
$TTCV_t = \sum_{i \in I} \sum_{z \in Z} \sum_{g \in Gv} \sum_{e \in V} \left(\alpha_t UTV_{igvz} QV_{igvzt} \right), \quad \forall \, t$ is the transportation cost for grain for food securitywhere

$$\left. \begin{aligned} UTC_{igfl} &= IA_{il} + \left(IB_{il} ADG_{gfl} \right) \\ UTCs_{igfl} &= OAU_e + \left(OBU_e ADG_{gfl} \right) \\ UTB_{fcb} &= OA_b + \left(OB_b ADF_{fcb} \right) \\ UTD_{dcb} &= OAD_b + \left(OBD_b ADD_{dcb} \right) \\ UTW_{fwm} &= OAW_{em} + \left(OBW_m ADW_{fwm} \right) \\ UTU_{gue} &= OAU_e + \left(OBU_e ADU_{gue} \right) \\ UTV_{igvz} &= OAV_{iz} + \left(OBV_{iz} ADV_{gvz} \right) \end{aligned} \right\},$$

IA_{il} and IB_{il} are fixed and variable cost for transportation biomass type $i \in I$, (OA_b, OB_b) are fixed and variable cost for transportation bioethanol (E100), OAD_b and OBD_b are fixed and variable cost for transportation gasoline, OAW_m and OBW_m are fixed and variable cost for transportation solid waste, OAU_e and OBU_e are fixed and variable cost for transportation slow, and OAV_{iz} and OBV_{iz} are fixed and variable cost for transportation biomass type $i \in I$.

The biomass transportation cost UTC_{igfl} is described by Börjesson and Gustavsson, [25] for transportation by tractor, truck, and train UTB_{fcb}. They are composed of a fixed cost (IA_{il}, OA_b) and a variable cost (IB_{il}, OB_b). Fixed costs include loading and unloading costs. They do not depend on the distance of transport. Variable costs include fuel cost, driver cost, maintenance cost, etc. They are dependent on the distance of transport [26].

A Carbon Tax Levied Cost Model $TTAXB_t$

Many countries are implementing various mechanisms to reduce GHG emissions including incentives or mandatory targets to reduce carbon footprint. Carbon taxes and carbon markets (emissions trading) are recognized as the most cost-effective mechanisms. The basic idea is to put a price tag on carbon emissions and create new investment opportunities to generate a fund for green technology development. There are already a number of active carbon markets for GHG emissions [27, 28].

A carbon tax levied is determined by the equation:

$$TTAXB_t = (\alpha_t TEI_t) C_{CO_2}, \quad \forall\ t \tag{4.18}$$

Government Incentives for Bioethanol (E100) Production Cost Model

Government incentives TL_t for bioethanol (E100) production and use is determined by the equation:

$$TL_t = \sum_{f \in F} \sum_{c \in C} \sum_{b \in B} (INS_{ft} \alpha_t QB_{fcbt}), \quad \forall\ t \tag{4.19}$$

Total Costs of Selling Straw for Other Purposes

$$TS_t = \alpha_t \sum_{i \in I} \left(PSU_{it} \sum_{g \in G} \sum_{u \in U} \sum_{e \in E} QU_{iguet} \right), \quad \forall\ t \tag{4.20}$$

5.3.3   Model of Social Assessment of an IBSC Job_t, [*Number of Jobs/year*]

The IBSC Social Assessment Model (see Eq. 4.20) is to determine the expected **total number of jobs created** (Job_t) as a result of the operation of all elements of the system during its operation.

$$Job_t = NJ1_t + LT_t NJ2_t + LT_t NJ3_t, \quad \forall \ t \tag{4.21}$$

where the components of (4.20) are defined according to the relations for each time interval $t \in T$, [*Number of Jobs/year*]:

$NJ1_t$ number of jobs created during the installation of bioethanol refineries and solid
  waste plants,
$NJ2_t$ number of jobs created during the operation of bioethanol refineries and solid
  waste plants,
$NJ3_t$ number of jobs created by cultivation bioresources for bioethanol production,
and are determined according to the equations:

$$\left.\begin{array}{l} NJ1_t = \sum\limits_{p \in P}\sum\limits_{f \in F} \left(M_{ft}^{JobP} Job\, B_p\, Z_{pft} \right) + \sum\limits_{s \in Sw}\sum\limits_{w \in W} \left(M_{wt}^{JobW} Job\, W_s\, ZW_{swt} \right) \\[2ex] NJ2_t = \sum\limits_{p \in P}\sum\limits_{f \in F} \left(M_{ft}^{JobP} Job\, O_p\, ZF_{pft} \right) + \sum\limits_{s \in Sw}\sum\limits_{w \in W} \left(M_{wt}^{JobW} Job\, OW_s\, ZWF_{swt} \right) \\[2ex] NJ3_t = \sum\limits_{i \in I}\sum\limits_{g \in G} \left(Job\, G_{ig} PBB_{igt} \right) \end{array}\right\}, \quad \forall \ t$$

$$\tag{4.22}$$

Equations (4.21) and (4.22) represent a simplified model of assessment of the social criterion, which was first discussed in detail in [29].

5.4   Restrictions

5.4.1   SC Design Constraints

These constraints are material balances among the different nodes in the SC. Following, the constraints between different SC nodes are formulated:

Bioethanol Plants Capacity Limited by Upper and Lower Constrains

Plants capacity is limited by upper and lower bounds, where the minimal production level in each region is obtained by

$$\sum_{p \in P} \left(PB_p^{MIN} ZF_{pft} \right) \leq \alpha_t \sum_{c \in C} \sum_{b \in B} QB_{fcbt} \leq \sum_{p \in P} \left(PB_p^{MAX} ZF_{pft} \right), \quad \forall f,t \qquad (4.23)$$

$$\left(PBS_{ip}^{MIN} ZF_{pft} \right) \leq \left(\alpha_t \sum_{c \in C} \sum_{b \in B} QBP_{ifcbpt} \right) \leq \left(PBS_{ip}^{MAX} ZF_{pft} \right), \quad \forall i,f,p,t \quad (4.24)$$

$$\left(PBSs_{ip}^{MIN} ZF_{pft} \right) \leq \left(\alpha_t \sum_{c \in C} \sum_{b \in B} QBPs_{ifcbpt} \right)$$

$$\leq \left(PBSs_{ip}^{MAX} ZF_{pft} \right), \quad \forall i,f,p,t \qquad (4.25)$$

where

$$QB_{fcbt} = \sum_{i \in I} \sum_{p \in P} \left(QBP_{ifcbpt} + QBPs_{ifcbpt} \right), \quad \forall t,f,c,b$$

Solid Waste Plants Capacity Limited by Upper and Lower Constrains

A condition that ensures that the total amount of solid waste generated by all biorefineries can be processed in the plants built for this purpose

$$\sum_{w \in W} \sum_{m \in M} QW_{fwmt} \leq \sum_{p \in P} \sum_{i \in I} \sum_{c \in C} \sum_{b \in B} \left(SW_{ipt} QBP_{ifcbpt} + SWs_{ipt} QBPs_{ifcbpt} \right), \quad \forall f,t$$

$$(4.26)$$

$$\left. \sum_{s \in S} \left(P_s^{MIN} ZWF_{swt} \right) \leq \alpha_t \sum_{f \in F} \sum_{m \in M} QW_{fwmt} \leq \sum_{s \in S} \left(P_s^{MAX} ZWF_{swt} \right) \right\}, \quad \forall t \qquad (4.27)$$

$$QW_{fwmt} \leq \sum_{s \in S} \left(P_s^{MAX} ZWF_{swt} \right), \quad \forall t,w,f,m, \qquad (4.28)$$

Limits on IBSC Flow Acceptability

Equation (4.29) guarantees the permissible values of grain and straw flows from each of the regions for their production

$$\left.\begin{aligned} \sum_{l\in L}\sum_{f\in F}QI_{igflt} + \sum_{z\in Z}\sum_{v\in V}QV_{igvzt} \le QI_{igt}^{MAX} \\ \sum_{e\in E}\sum_{f\in F}QIs_{igfet} + \sum_{e\in E}\sum_{u\in U}QU_{iguet} \le QIs_{igt}^{MAX} \end{aligned}\right\} \quad \forall t,g,i \qquad (4.29)$$

Equation (4.30) guarantees the admissibility of solid waste streams for each of the places where they are generated

$$\sum_{m\in M}\sum_{w\in W}QW_{fwmt} \le QW_{ft}^{MAX}, \quad \forall m,f \qquad (4.30)$$

Equation (4.31) guarantees the permissible values of bioethanol flows from each of the regions for their production

$$\left.\begin{aligned} \sum_{i\in I}\sum_{c\in C}\sum_{b\in B}\left(QBP_{ifcbpt} + QBPs_{ifcbpt}\right) \le QB_{ft}^{MAX}ZF_{pft} \\ \sum_{b\in B}\sum_{c\in C}\left(QB_{fcbt}\right) \le QB_{ft}^{MAX}ZF_{pft} \end{aligned}\right\}, \quad \forall\ t,f,p \qquad (4.31)$$

Constraints balance of bioethanol (E100) to be produced from biomass available in the regions:

$$G^{MAX}ZF_{pft} \ge QBP_{ifcbpt}, \quad \forall\ i,f,c,b,p,t \qquad (4.32)$$

In performing (4.33), we can write the expression for

$$QB_{fcbt} = \sum_{i\in I}\sum_{p\in P}QBP_{ifcbpt}, \quad \forall\ t,f,c,b \qquad (4.33)$$

Equation (4.34) guarantees the permissible values of gasoline flows from each of the regions for their production

$$\sum_{b\in B}\sum_{c\in C}\left(QD_{dcbt}\right) \le QD_{dt}^{MAX}, \quad \forall\ t,d \qquad (4.34)$$

Equations (4.35) and (4.36) guarantee the permissible values of the flows of grain and straw from each of the regions for their production

$$\left.\begin{array}{l}\dfrac{\left(A_{igt}+A_{igt}^{F}\right)}{\beta_{igt}\alpha_{t}}\geq\left(\displaystyle\sum_{l\in L}\sum_{f\in F}QI_{igflt}+\sum_{z\in Z}\sum_{v\in V}QV_{igvzt}\right)\\[4mm]\dfrac{\left(A_{igt}+A_{igt}^{F}\right)}{\beta s_{igt}\alpha_{t}}\geq\left(\displaystyle\sum_{e\in E}\sum_{f\in F}QIs_{igfet}+\sum_{e\in E}\sum_{u\in U}QU_{iguet}\right)\end{array}\right\},\quad \forall\ t,g,i \qquad (4.35)$$

$$\left.\begin{array}{l}\displaystyle\sum_{l\in L}\sum_{g\in G}QI_{igflt}=\sum_{c\in C}\sum_{b\in Bp}\sum_{\in P}\left(\dfrac{QBP_{ifcbpt}}{\gamma_{ipt}}\right)\\[4mm]\displaystyle\sum_{l\in L}\sum_{g\in G}QIs_{igflt}=\sum_{c\in C}\sum_{b\in Bp}\sum_{\in P}\left(\dfrac{QBPs_{ifcbpt}}{\gamma s_{ipt}}\right)\end{array}\right\},\quad \forall\ t,f,i \qquad (4.36)$$

A Limitation Guaranteeing the Regions Needs for Straw for Technical Needs and Utilization

$$PSTRAW_{iut}^{MIN}\leq\alpha_{t}\sum_{e\in E}\sum_{g\in G}QU_{iguet}\leq PSTRAW_{iut}^{MAX},\quad \forall\ t,i,u \qquad (4.37)$$

A Limitation Guaranteeing the Regions Needs for Grain for Technical Needs and Utilization

$$\alpha_{t}\sum_{z\in Z}\sum_{g\in G}QV_{igvzt}=PGRAIN_{ivt},\quad \forall\ t,i,v \qquad (4.38)$$

Mass Balances Between Bioethanol (E100) Plants and Biomass Regions

The connections between bioethanol (E100) plants and biomass regions are given by

$$\left.\begin{array}{l}\displaystyle\sum_{l\in L}\sum_{g\in G}QI_{igflt}\leq\sum_{p\in P}\left(\dfrac{PBS_{ip}^{MAX}ZF_{pft}}{\gamma_{ipt}}\right)\\[4mm]\displaystyle\sum_{l\in L}\sum_{g\in G}QIs_{igflt}\leq\sum_{p\in P}\left(\dfrac{PBSs_{ip}^{MAX}ZF_{pft}}{\gamma s_{ipt}}\right)\end{array}\right\},\quad \forall\ t,f,i \qquad (4.39)$$

Mass Balances Between Bioethanol (E100) Plants and Customer Zones

$$\sum_{b\in B}\sum_{f\in F}\left(\alpha_t QB_{fcbt}\right) = QEB_{ct}, \quad \forall\ t,c \tag{4.40}$$

5.4.2 Logical Constrains

These constraints define the structure of the IBSC and are as follows:

Restriction Guarantees That a Given Region $f \in F$ Installed Power Plant with Size $p \in P$ for Bioethanol (E100) Production

Constraint (4.41) states that can be chosen only one size $p \in P$ for each facility.

$$\left.\begin{array}{c}\sum_{p\in P}Z_{pft} \leq 1 \\[2mm] \sum_{p\in P}ZF_{pft} \leq 1\end{array}\right\}, \quad \forall\ t,f \tag{4.41}$$

Restriction Guarantees That a Given Region $w \in W$ Installed Solid Waste Plant with Size $s \in S$

$$\left.\begin{array}{c}\sum_{s\in S}ZW_{swt} \leq 1 \\[2mm] \sum_{s\in S}ZWF_{swt} \leq 1\end{array}\right\}, \quad \forall\ t,w \tag{4.42}$$

Constraint (4.42) states that can be chosen only one size $s \in S$ for each utilization plant.

Limitation Ensure the Availability of at Least One Connection to a Region of Bioresources and Region for Biofuel

$$\sum_{g\in G}\sum_{l\in L}X_{igflt} \geq \sum_{c\in C}\sum_{b\in B}Y_{fcbt} \geq \sum_{p\in P}ZF_{pft}, \quad \forall\ t,i,f \tag{4.43}$$

Limit Which Guarantees That Each Region Will Provide Only One Plant
with a Biomass Type $i \in I$

$$\sum_{f \in F} \sum_{l \in L} X_{igflt} \leq 1, \quad \forall \ t, i, g \tag{4.44}$$

Limitation of Assurance That At Least One Region $f \in F$ Producing
Bioethanol (E100) Is Connected to a Costumer Zones $c \in C$

$$\sum_{b \in B} \sum_{f \in F} Y_{fcbt} \leq 1, \quad \forall \ t, c \tag{4.45}$$

Limitation of Assurance That At Least One Region $f \in F$ Producing
Bioethanol (E100) Is Connected to a Solid Waste Utilization Plant Located
in Region $w \in W$

$$\sum_{w \in W} \sum_{m \in M} WS_{fwmt} \leq 1, \quad \forall \ t, f \tag{4.46}$$

Condition Ensuring That the Solid Waste Produced from a Given Biorefinery
Will Be Processed in Only One of the Plants for Use

$$\sum_{m \in M} \sum_{w \in W} WS_{fwmt} = \sum_{p \in P} ZF_{pft}, \quad \forall \ t, f \tag{4.47}$$

Condition Ensuring That a Plant Used in a Given Region Will Be Connected
to At Least One Plant in Which Solid Waste Is Generated

$$\sum_{m \in M} \sum_{f \in F} WS_{fwmt} \geq \sum_{s \in S} ZWF_{swt}, \quad \forall \ t, w \tag{4.48}$$

**Restrictions That Ensure That Only One Mode of Transport Is Used
for the Transport of Bioraw Materials or Finished Products Between Two
Areas**

1. Transporting the grain to produce bioethanol from a region $g \in G$ to $f \in F$ in a
   given time interval is carried out by only one type of transport,

$$\sum_{l \in L} X_{igflt} \leq 1, \quad \forall \ t, i, g, f \tag{4.49}$$

2. The transport of straw to produce bioethanol from a region $g \in G$ to $f \in F$ in a
   given time interval is carried out by only one type of transport,

$$\sum_{l \in L} Xs_{igflt} \leq 1, \quad \forall \ t, i, g, f \tag{4.50}$$

3. Bioethanol transportation from a region $f \in F$ to $c \in C$ in a given time interval is
   carried out by only one type of transport,

$$\sum_{b \in B} Y_{fcbt} \leq 1, \quad \forall \ t, f, c \tag{4.51}$$

4. The transportation of solid waste from a region $f \in F$ to $w \in W$ in a given time
   interval is carried out by only one type of transport,

$$\sum_{m \in M} WS_{fwmt} \leq 1, \quad \forall \ t, f, w \tag{4.52}$$

5. The transportation of straw for processing and utilization from the region $g \in G$ to
   $u \in U$ in a given time interval is carried out by only one type of transport,

$$\sum_{e \in E} WU_{guet} \leq 1, \quad \forall \ t, g, u \tag{4.53}$$

6. The transportation of grain for food purposes from a region $g \in G$ to $v \in V$ in a
   given time interval is carried out by only one type of transport,

$$\sum_{z \in Z} WV_{gvzt} \leq 1, \quad \forall \ t, g, v \tag{4.54}$$

7. Gasoline transportation from a region $d \in D$ to $c \in C$ in a given time interval is
   carried out by only one type of transport,

$$\sum_{b \in B} DT_{dcbt} \leq 1, \quad \forall \ t, d, c \tag{4.55}$$

5.4.3 Transport Links

Restrictions on Transportation of Biomass Are

$$
\left.
\begin{aligned}
PBI_{ig}^{MIN}\sum_{l\in L}X_{igflt} \le \alpha_t\sum_{l\in L}QI_{igflt} \le PBI_{ig}^{MAX}\sum_{l\in L}X_{igflt}\\
PBIS_{ig}^{MIN}\sum_{l\in L}X_{igflt} \le \alpha_t\sum_{e\in E}QIs_{igfet} \le PBIS_{ig}^{MAX}\sum_{l\in L}X_{igflt}
\end{aligned}
\right\},\quad \forall\ t,i,g,f \qquad (4.56)
$$

Restrictions on Transportation of Bioethanol (E100) Are

$$
\sum_{b\in B}QB_{fcbt} \le G^{MAX}\sum_{b\in B}Y_{fcbt}, \quad \forall\ t,f,c \qquad (4.57)
$$

Restrictions on Transportation of Solid Waste Are

$$
\sum_{m\in M}QW_{fwmt} \le G^{MAX}\sum_{m\in M}WS_{fwmt}, \quad \forall\ t,f,w \qquad (4.58)
$$

Restrictions on Transportation of Straw Are

$$
\sum_{e\in E}QU_{guet} \le G^{MAX}\sum_{e\in E}WU_{guet}, \quad \forall\ t,g,u \qquad (4.59)
$$

Restrictions on Transportation of Wheat–Corn for Food Security Are

$$
\sum_{z\in Z}QV_{gvzt} \le G^{MAX}\sum_{z\in Z}WV_{gvzt}, \quad \forall\ t,g,v \qquad (4.60)
$$

5.4.4 Restriction for Total Environmental Impact of All Regions

$$TEI_t \leq TEI_t^{MAX}, \quad \forall\ t \tag{4.61}$$

where TEI_t^{MAX} is the maximum permissible values for the total environmental impact of IBSC and fossil fuel in the regions for each time interval $t \in T$, $[kg\, CO_2 - eq\, d^{-1}]$.

5.4.5 Limitation Guaranteeing Crop Rotation

The crop rotation allows ensuring control of pests, improving soil fertility, maintenance of the long-term productivity of the land, and consequently increasing the yields and profitability of the rotation [30, 31]. The combination of crop rotation and fallowing is a common practice that is gaining momentum again due to environmental benefits and promoted reduction in the dependence on external inputs.

Crop rotation implemented in a region $g \in G$ means that the growing area of energy crops are rotated so that the next time the same area is used by other crops grown under are optimal scheme of crop rotation. This can be achieved if for land A_{igt} and A_{igt}^F inequalities are implemented:

$$\left(A_{igt} + A_{igt}^F\right)2.0 \leq \left(A_{gt}^S - A_{gt}^{Food}\right), \quad \forall\ i, g, t \tag{4.62}$$

5.4.6 Model of Constraints for Energy Balances/Energy Efficiency Constraints

Limitation Ensuring That the Overall Energy Balance in the Region Is Provided

The demand for bioethanol in each of our regions is accepted in proportion to the population, taking into account the percentage of bioethanol in the gasoline blend for the relevant time horizon.

Limitation of enforceability of the energy balance:

$$EGD_t + EB_t \geq EO_t, \quad \forall\ t. \tag{4.63}$$

Energy equivalent gasoline, which is necessary to meet the energy needs of the all customer zones where no use bioethanol (E100) is determined by the equation:

$$EO_t = ENO \sum_{c \in C} YO_{ct}, \quad \forall\ t, \tag{4.64}$$

where EO_t is annual requirement of energy (gasoline) of all regions $[GJ\, year^{-1}]$.

In cases where the fuel consumption is considered proportional to the population of a given region, then the YO_{ct} is calculated according to the dependence:

$$YO_{ct} = \frac{PC_{ct}}{\sum_{c \in C} PC_{ct}} YG_t$$

The energy equivalent of petroleum diesel that must be added, in order to balance the energy required for all customer zones is determined by the equation:

$$EGD_t = ENO \sum_{c \in C} QEG_{ct}, \quad \forall \ t, \tag{4.65}$$

where $EGD_t \ [GJ \, year^{-1}]$ is annual energy added to gasoline fuel to balance the required energy for all regions and $QEG_{ct} = \alpha_t \sum_{b \in B} \sum_{d \in D} QG_{dcbt} \ [ton/year]$.

The energy equivalent of bioethanol (E100) received per year of work IBSC is determined according to the dependence:

$$EB_t = ENB \sum_{c \in C} QEB_{ct}, \quad \forall \ t, \tag{4.66}$$

where $EB_t \ [GJ \, year^{-1}]$ is annual energy received from the extracted biofuel (bioethanol(E100)) of IBSC for all customer zone and $QEB_{ct} =$

$\alpha_t \sum_{b \in B} \sum_{f \in F} QB_{fcbt} \ [ton/year]$.

Total cost of fuel used from the regions [$/year] is solved from equation:

$$TBG_t = TDC_t + PG_t \sum_{c \in C} QEG_{ct}, \quad \forall \ t \tag{4.67}$$

Limitation Ensuring That the Overall Energy Balance in Each Customer Zones Is Provided

Limitation of enforceability of the energy balance for each region:

$$ENO \ QEG_{ct} + ENB \ QEB_{ct} \geq ENO \ YO_{ct}, \quad \forall \ t, c. \tag{4.68}$$

Limitation Ensuring That Each Region Will Be Provided in the Desired Proportions Fuels

$$ENB\sum_{f\in F}\sum_{c\in C}\sum_{b\in B}(\alpha_t QB_{fcbt}) \geq K_{ct}^{mix}ENO\sum_{c\in C}YO_{ct}, \quad \forall \ t \tag{4.69}$$

$$ENB\sum_{f\in F}\sum_{b\in B}(\alpha_t QB_{fcbt}) \geq K_{ct}^{mix}ENO \ YO_{ct}, \quad \forall \ t,c \tag{4.70}$$

5.4.7   Model of Constraints for Total Cost of a BSC Network

$$TDC_t^{MAX} \geq TDC_t, \quad \forall \ t \in T \tag{4.71}$$

where TDC_t^{MAX} is the maximum total cost of a bioethanol (E100) SC network [$].

5.5   *Optimization Objective Functions*

As explained in [32], there is a vast literature on the research domain of supply chain design and management. There are also numerous papers dealing with location problems. Initially, the optimization of the supply chain was made to achieve cost saving. As a result, all the costs that have an influence on the supply chain performance have to be considered simultaneously. Besides cost considerations, more recently some papers have enlarged the system performance criteria by including energy consumption and GHG emissions across the SC as in *the Integrated Biomass Supply Analysis and Logistics* (IBSAL) model proposed by [33] for corn stover to biorefineries, or in the work of [34] for biofuel production. But to evaluate the global performance of a system, it is necessary to describe how human activity can impose different types of impacts on global sustainability, i.e., simultaneous progress in economic profitability, environment preservation, and social consideration. Thus the use of the multi-objective optimization method prior requires translating all the sustainable aspects into suitable criteria that could be optimized simultaneously. Till now the social assessment is often neglected. However, to the best of our knowledge, except the work of [35], no study integrates a complete sustainable development view by adding a suitable social criterion to both previous ones in order to optimize the supply chain of industrial products. The main reason is that the evaluation of the social indicators is often a tremendous and difficult task.

The model includes the following objectives:

5.5.1 Economic Objective Function

The part of the objective function associated with the minimization of the economic costs includes all the operating costs of the supply chain, from the purchase of biomass feedstock to transportation of the final product, as well as the investment cost of biorefineries and storage facilities. The costs of the supply chain are: the cost of raw material, the transport of raw material to the collection facilities, the cost of handling and storage of biomass, the cost of transport to the biorefineries, the cost of transformation into bioethanol (E100), and the cost of final transport to the blending facilities. The economic objective is to minimize the total annual costs. The terms of the cost objective corresponding to the annual operation costs of the supply chain (AOC) are described in the following equation:

$$COST = \sum_{t \in T}(LT_t TDC_t) \tag{4.72}$$

In Eq. (4.72), the summation terms represent, respectively, the annual operating costs for biomass cultivation and harvesting, biomass transport, biomass inventory, biomass conversion, ethanol inventory, and ethanol transport.

As a target function we can also use the price of the used fuel (petrol and bioethanol (E100)) for the entire time interval, provided that the needs of the regions of this energy carrier are satisfied. This is determined by the dependency:

$$COST_{TBG} = \sum_{t \in T}(LT_t TBG_t) \tag{4.73}$$

5.5.2 Environmental Objective Function

The environmental objective function corresponds to the minimization of the entire environmental impact measured through the Eco indicator 99 method. The Eco indicator 99 is a standard method for evaluating the global environmental impact of a process; product and/or activity [36]. This method can be applied either as a standalone tool or combined with an optimization model. The proposed model integrates the Eco indicator 99, whose calculation has been carried out considering the specific activities taking place in the operation of the considered IBSC. Particularly, this paper considers the damages to the ecosystem quality, human health, and resources.

The cumulative environmental impact of system performance, expressed by the amount of carbon dioxide equivalent generated over the whole life cycle and during its operation, is expressed by means of the equation:

$$ENV = \sum_{t \in T}(LT_t TEI_t) \tag{4.74}$$

The environmental impact [32] is quantified with the eco costs method introduced by [37, 38], and updated in 2007 and 2012. Eco costs are a measure that expresses the environmental load of a product on the basis of prevention of that burden during the product life cycle: from the raw materials until its end of life. This indicator represents the necessary costs that should be made to counteract the negative impact of the activity made on the capacity of earth [39]. It quantifies the impact in terms of pollution and material depletion by allocating a cost penalizing the use of an alternative that would reduce its impact on the environment and would be called sustainable solution.

Eco costs allow quantifying the environmental impact as a simple indicator easy to understand and compare with other criteria, for example, economic. Furthermore, as [38] has declared, the main advantages of these Eco costs are: (1) they are expressed as a monetary value, (2) there is no need to compare with another product (often the case with other life cycle assessment methods), and (3) calculations are based on European price levels and the costs are updated.

In our case, to determine the monetary equivalent of the environmental impact, we use the Global warming coefficient C_{CO_2} using the equation:

$$Cost_{ENV} = C_{CO_2} ENV \tag{4.75}$$

where $Cost_{ENV}[\$ \, year^{-1}]$ is the price to be paid to prevent the impact on the environment of the amount of carbon dioxide equivalent, while C_{CO_2} is the global warming coefficient $[\$/kg \, CO_2 - eq]$ (the most commonly used values is $0.135\$/kg \, CO_2 - eq$) according to [40].

5.5.3 Integrated Economic and Environmental Objective Function

The integrated economic and environmental objective function is formulated as follows:

$$Int\_COST = COST_{TBG} + C_{CO_2} \sum_{t \in T} (a_t LT_t TEI_t) \tag{4.76}$$

where
a_t
IBSC operating period for 1 year $[d/y]$;

The total emissions are converted into carbon credits by multiplying with the carbon price C_{CO_2} on the market, where it has a value $0.135\$/kg \, CO_2 - eq$ [40].

5.5.4 Social Objective Function

Till now the social assessment is often neglected. However, to the best of our knowledge, except the work of [35], no study integrates a complete sustainable development view by adding a suitable social criterion to both previous ones in order to optimize the supply chain of industrial products. The main reason is that the evaluation of the social indicators is often a tremendous and difficult task.

According to [32], the goal is to quantify the social sustainability of a system. In our approach as the system to implant is completely new, most of the social impacts would remain almost constant for instance human health and security risks or public acceptability. In our case the two major social indicators are the jobs creation and the food to energy one. The latter assesses the possible competition between food and energy. In [41] the authors have explained that it is used as a social indicator as it deals with measuring the quality of life: rise of prices of food, and threat of the safety of food supply. As one aim is to compare first and second generation of biorefineries, this indicator must be taken into account to clearly establish the discrepancy between grain and straw as a biomass. But this competition is already evaluated through the eco costs relative to land-use.

Concerning jobs estimation, the most important problems are to define the boundary of the evaluation and then to calculate the total number of jobs created. Indeed, this number is not limited to the number of persons who are directly working for the new activity but it must also take into account the jobs created or supported by subcontractors and more generally by all the firms impacted in terms of employments. As a consequence the number of jobs created is classically divided into three categories: (1) Direct jobs (jobs related to plant's operations), (2) Indirect jobs (new employees in subcontractors), and (3) Induced jobs (new employees in the local economy). This last number evaluates the employments generated by the two previous categories due to their (and their families) consumption in the local economy.

For an estimate of the social impact of the system work, we can use the exacted coefficients of ($JobB_p$, $JobO_p$, $JobOW_s$, $JobW_s$) which account for indirect jobs in the local economy. Then the social impact (in terms of jobs) is determined according to the dependence [*Number of Jobs*]:

$$JOB = \sum_{t \in T}(LT_t Job_t) \qquad (4.77)$$

6   Optimal Synthesis Problem Formulation Using Mathematical Model

The optimization procedure finds the set of decision variables, both binary and continuous, that minimize of the objective function. The identified decision variables are:

- SC network structure, which includes: number, size, and location of biorefineries,
- biomass cultivation rate for each biomass feedstock type and bioethanol(E100) production,
- locations of bioethanol(E100) production facilities and biomass cultivation sites,
- flows of each biomass type and bioethanol(E100) between cells,
- modes of transport of delivery for biomass and bioethanol(E100),
- greenhouse gas emissions for each stage of the life cycle,
- transportation amount for each transportation link and transportation mode,
- distribution processes for biofuel to be sent to mixing and demand zones.

In the following model, two objective functions are considered:

- **Economic sustainability** (*COST* or *COST_{TBG}*) (4.72, 4.73): Minimize the total logistics cost of the supply chain considering fixed, variable, and emissions costs [$].
- **Environmental sustainability** (*ENV* or *Cost_{ENV}*) (4.74, 4.75): Minimize the total quantity of GHG emissions calculated in units of [*kg* or $] of carbon dioxide equivalent [$kg\, CO_2 - eq$].
- **Social sustainability** (*JOB*) (4.76): Maximize the social impact of the system work of the supply chain [*Number of Jobs*].

The problem for the optimal design of an IBSC is formulated as a MILP model for different target functions as follows:

The problem for the optimal design of a IBSC is formulated as a MILP model Since all constraints are linear functions of the continuous and binary variables, and the formulated objective functions are linear.

6.1 Single-Criteria Objective Models

The first approach considers that the SCM problem has only one objective function to optimize which usually represents the economic or environmental dimension.

Strategic SC design integrates two planning levels: decisions on the SC network configuration and the mission of each refinery and planning decisions on the flows of biomass and fuels in the network.

6.1.1 Minimizing GHG Emissions [$kg\, CO_2 - eq$]

As discussed in Sect. 6.1.1 environmental objective is to minimize the total annual CO_2-equivalent GHG emissions resulting from the operations of the IBSC and gasoline used to provide the energy balance of the regions. The formulation of this objective is based on total GHG emissions in the IBSC and other fuels are estimated based on LCA approach, where emissions are added every life stage.

The task of determining the optimal location of facilities in the regions and their parameters is formulated as follows:

$$\left\{ \begin{array}{l} Find : X_t [\text{Decision variables}]^T \\ MINIMIZE\{ENV(X_t)\} \rightarrow (\text{Eq.4.73}) \\ s.t. : \{\text{Eq.4.22} - \text{Eq.4.70}\} \end{array} \right\} \qquad (4.78)$$

The objective function (Eq. 4.73) and restrictions by (Eq. 4.22)–(Eq. 4.70) are linear with respect to all decision variables.

6.1.2 Minimizing Annualized Total Cost [$]

The single objective based models consider that the economic dimension is the most important and there is no need to integrate other objective such as the environmental objective (GHG emissions).

The economic objective is to minimize the annualized total cost, including the total annualized capital cost, the annual operation cost, the annual governmental incentive, and the cost for emitting CO_2. The task of determining the optimal location of facilities in the regions and their parameters is formulated as follows:

$$\left\{ \begin{array}{l} Find : X_t [\text{Decision variables}]^T \\ MINIMIZE\{COST(X_t)\} \rightarrow (\text{Eq.4.71}) \\ s.t. : \{\text{Eq.4.22} - \text{Eq.4.70}\} \end{array} \right\} \qquad (4.79)$$

The objective function (Eq. 4.71) and restrictions by (Eq. 4.22)–(Eq. 4.70) are linear with respect to all decision variables.

6.1.3 Maximize the Social Impact of the System Work of the Supply Chain

$$\left\{ \begin{array}{l} Find : X_t [\text{Decision variables}]^T \\ MAXIMIZE\{JOB(X_t)\} \rightarrow (\text{Eq.4.75}) \\ s.t. : \{\text{Eq.4.22} - \text{Eq.4.70}\} \end{array} \right\}, \quad \forall \ t \in T \qquad (4.80)$$

The objective function (Eq. 4.75) and restrictions by (Eq. 4.22)–(Eq. 4.70) are linear with respect to all decision variables.

6.2   Multi-Criteria Objective Models

The second class formulates strategic SC decisions as multi-criteria/multi-objective programs. The planning decisions are almost the same. However, additional objectives are added in the optimization process.

Sustainable SC management covers interactions among economic dimension, the environment, and society, and a realistic decision process should find a trade-off solution between different performances which are sometimes conflicting. Thus, the use of multi-criteria and multi-objective models is suitable in this case.

In this context, Multi-Objective Optimization (MOO) models provide decision makers with the possibility to understand the trade-off between different objectives and their impact on the SC configuration and planning decisions and costs. Different methods for solving the MOO problem could be used such as reference point methods with weight coefficients, ε-constraint, and goal programming (GP). Although multi-objective optimization might add another degree of complexity to the decision process, especially when the decision makers have to give their preferences (weight for objectives), it is more representative to the real life strategic planning process.

The reference point approach is using target values F_{ti}^{ref} found by solving the optimization problem separately for a given set of objectives (Eq. 4.43)–(Eq. 4.46) and finding a Pareto optimal solution minimizing the generalized function of losses from the reference values F_{ti}^{ref}. For example of two objectives—total costs TDC_t and TEI_t GHG emissions, the decision making problem is solved as following:

$$
\left(
\begin{array}{l}
Find : X_t[\text{Decision variables}]^T \\[4pt]
To\ MIN.: F_t(X_t) = \left(
\begin{array}{l}
TDC_t = Total\ \cos t \\[4pt]
TEI_t = \text{GHG Emission}
\end{array}
\right) \\[4pt]
s.t. : \{\text{Eq.4.19} - \text{Eq.4.42}\}
\end{array}
\right)
$$

$$
\Rightarrow
\left(
\begin{array}{l}
Find : X_t[\text{Decision variables}]^T \\[4pt]
To\ MIN. : Z_t(X_t) = \sum_{i=1}^{2} \left(\delta_{ti}^2(X_t)w_{ti}\right) \\[4pt]
s.t. : \{\text{Eq.4.19} - \text{Eq.4.42}\}
\end{array}
\right)
\qquad (4.81)
$$

where $\delta_{ti}(X_t)$ are the normalized losses for each objective from the reference value F_{ti}^{ref} and w_{ti} are the weight coefficients, representing the priorities given to each objective.

The method of "ε-constraint" is used to minimize costs while ensuring eligibility of greenhouse gas emissions throughout the life cycle. In this method (with posterior articulation of preference), one objective is selected for optimization and the others are reformulated as constraints, i.e.:

$$\begin{pmatrix} Find : X_t[Decision\ variables]^T \\ To\ MIN. : F_t(X) = \begin{pmatrix} TDC_t = Total\ \cos t\ IBSC \\ TEI_t = GHG\ Emissions \end{pmatrix} \\ s.t. : \{Eq.4.19 - Eq.4.42\} \end{pmatrix}$$
$$\Rightarrow \begin{pmatrix} Find : X_t[Decision\ variables]^T \\ To\ MIN. : F_t(X_t) = TDC_t \\ s.t. : \begin{cases} TEI_t \leq \varepsilon \\ Eq.4.19 - Eq.4.42 \end{cases} \end{pmatrix} \qquad (4.82)$$

By progressively changing the constraint values ε, which represent the limit on GHG emissions in this case, different points on the Pareto-front could be sampled. By calculating the extremes of the Pareto-front the range of different objective functions could be calculated and constraint values selected accordingly.

The second method is the goal programming (GP). The GP model could be placed in the third category. The algebraic formulation of GP is given as following:

$$\begin{pmatrix} Find : X_t[Decision\ variables]^T \\ To\ MIN. : F_t(X_t) = \begin{pmatrix} TDC_t = Total\ \cos t \\ TEI_t = GHG\ Emissions \end{pmatrix} \\ s.t. : \{Eq.4.19 - Eq.4.42\} \end{pmatrix}$$
$$\Rightarrow \begin{pmatrix} FinD : X_t[Decision\ variables]^T \\ To\ MIN. : Z_t = \sum_{i=1}^{2} (u_{ti}n_{ti} + v_{ti}p_{ti}) \\ s.t. : \begin{cases} TDC_t + n_{t1} + p_{t1} = F_{t1}^* \\ TEI_t + n_{t2} + p_{t2} = F_{t2}^* \\ Eq.4.19 - Eq.4.42 \end{cases} \end{pmatrix} \qquad (4.83)$$

where F_{t1}^*, F_{t2}^* is the target value for the objective function $F_t(X_t)$ which usually represents the minimum value obtained by considering this objective in the optimization process, n_{t1}, n_{t2} and p_{t1}, p_{t2} represent the negative and positive deviations from this target value for each time interval. The manager must analyze each one of the goals considered in the model in terms of whether over or underachievement of the goal is satisfactory where achievement implies that a goal has been reached. The terms u_{t1}, u_{t2} and v_{t1}, v_{t2} are the respective positive weights attached to these deviations in the achievement function Z_t. The weight factor of a given objective represents two different roles [42]. The first one is "normalization" that brings all deviations to a common unit of measurement. The second is "valorization" reflecting the decision maker's preference structure. For instance, these weights take the value zero if the minimization of the corresponding deviational variable is unimportant to decision makers. The "ε-constraint" method first helps the decision maker to identify different possible solutions and the characteristic of each objective. Once he obtains, he can go through a decision process where he can articulate the preference structure and choose the trade-off solution that guarantee the different objectives.

Sections 6.1 and 6.2 are ordinary MILP and can thus be solved using standard MILP techniques. The model was developed in the commercial software General Algebraic Modeling System (GAMS) [43]. The model will choose the less costly pathways from one set of biomass supply points to a specific plant and further to a set of bioethanol(E100) demand points. The final result of the optimization problem would then be a set of plants together with their corresponding biomass, bioethanol (E100), and gasoline demand points.

7 Optimal Renovation Problem Formulation Using Mathematical Model

The problem of optimal renovation of IBSC can be considered analogously to the problem of optimal synthesis as the mathematical models derived in Sect. 5, dependences for the used objective functions in Sect. 5.5 and the formulations for optimal synthesis in Sect. 6 are used. In the optimal renovation, it is assumed that the pre-existing system can retain all or part of the configuration of biorefineries and solid waste disposal plants, provided that the IBSC after the renovation has parameters satisfying the selected criterion (criteria). Under this assumption, the sets and basic parameters introduced in the synthesis problem are considered here to be composed of subsets before and after the renovation as shown below, respectively, for the set of factory plant sizes:

P is a unified set composed of the elements of subsets (SP, RP) and is represented as a combined set for which to write $P \equiv SP \cup RP$, i.e., $SP \subset P$ and $RP \subset P$, and none of the elements of the SP coincides with RP, i.e., $SP \notin RP$, where:

SP is the set of standard sizes from the plants that will be used in the process of renovation of IBSC, indexed by sp,

RP is the set of existing standard sizes from the plants that were used in IBSC before the renovation, indexed by rp.

Accordingly, for the sets of regions in which new bioethanol production installations are or may be additionally installed the set:

F is a combined set composed of the elements of the two sets (SF, RF) and is
  represented as a set for which to write $F \equiv SF \cup RF$, i.e., $SF \subset F$ and $RF \subset F$,
  and none of the elements of the SF coincides with RF, i.e., $SF \notin RF$, where:
SF is the many regions in which new bioethanol plants can be installed as a result of
  the renovation of the IBSC, indexed by sf
RF Represents the many existing bioethanol plants that were used in IBSC before the
  renovation, indexed by rf

Finally, for the many regions where solid waste treatment plants have been or will
be installed, the set:

S is a unified set composed of the elements of the two sets (SS, RS) and is
  represented as a set for which to write $S \equiv SS \cup RS$, i.e., $SS \subset S$ and $RS \subset S$
  and none of the elements of the SS coincides with RS, i.e., $SS \notin RS$, respectively.
SS is the set of standard sizes from the solid waste treatment plants that will be used
  in the IBSC renovation process, indexed by ss
RS is the set of existing sizes of solid waste treatment plants that were used in the
  IBSC before the renovation, indexed by rs.

Accordingly, for the sets of regions in which new solid waste disposal installations have been or may be additionally installed the set:

W is a unified set of regions composed of the elements of the two sets (SW, RW) and
  is represented as a set for which to write $W \equiv SW \cup RW$, i.e., $SW \subset W$ and
  $RW \subset W$, and none of the elements of the SW coincides with RW, i.e., $SW \notin RW$,
  where:
SW Represents the many regions in which new solid waste disposal plants can be
  installed as a result of the renovation of IBSC, indexed by sw
RW Represents the many regions where the existing solid waste disposal plants in
  the IBSC were installed before the renovation, indexed by rw

With the set P thus introduced above, we also define the new values of the
elements of $Cos\,B_p[\$]$ which correspond to the $\forall p \in RP$ are assumed to be equal
to "0" as they correspond to the biorefineries installed before the renovation (for
which the investment costs have already been paid) and be changed in the future. All
other elements of $Cos\,B_p[\$]$, that reply to the $\forall p \in SP$ occupy values corresponding
to the price of each biorefinery, which can be used in the renovation process and
generally depends on its size and depends on the technology used.

The values of the elements of $Cos\,W_s[\$]$ that reply to the $\forall s \in RW$ are assumed to
be equal to "0" as they correspond to the pre-renovation installations for solid waste
processing (for which the investment costs have already been paid) and will not be
changed in the future. For the other elements of $Cos\,W_s[\$]$ that correspond to ESW_t
have the values corresponding to the price of the newly built installation depending

on the size. Maximum annual plant capacity PB_p^{MAX} of size $p \in P$ for bioethanol (E100) production using grain and straw biomass type $i \in I$, for all $\forall p \in RP$ represent the values of the existing maximum capacities of which were used in the existing biorefineries before the renovation. Respectively for PB_p^{MIN}, PBS_{ip}^{MAX}, PBS_{ip}^{MIN}, $PBSs_{ip}^{MAX}$, and $PBSs_{ip}^{MIN}$ the same conditions apply.

The values of the elements of $JobB_p$ which represent the number of jobs occupied in the construction of organic plants of standard size P are formed as follows for all elements for $\forall p \in RP$, i.e., for the standard sizes of the biorefineries already built before the renovation, these values are "0," i.e., $JobB_p = 0$, $\forall p \in RP$ and therefore $JobB_p \neq 0$, $\forall p \in SP$ and equal to the actual jobs that are needed for the construction of a plant of the appropriate size in the conditions of renovation. Similarly for the elements of $JobW_s$ and "0" values are set for all $\forall s \in RS$, i.e., $JobW_s = 0$, $\forall s \in RS$ and therefore $JobW_s \neq 0$, $\forall s \in SS$ and equal to the actual number of jobs required to build a solid waste processing plant $\forall s \in SS$.

8   Potential Bioethanol Production in Bulgaria for 2016–2020

A case study of Bulgaria has been studied as an application of the proposed model. This case study examines an IBSC in the Bulgarian territory. Two major types of biomass resources, wheat and corn for production of first generation and wheat straw and corn cobs of second generation bioethanol are used. The proposed MILP model will determine the optimal level of the key logistics decision variables that maximize the expected profit of the IBSC.

8.1   Model Input Data

Bulgaria has 27 regions. In this case study, each region is considered to be a feedstock production region, a potential location of a biorefinery facility and a demand zone. In other words, the biofuel supply chain network consists of 27 areas for feedstock production, 27 potential biorefinery locations, 27 demand zones, 4 potential solid waste utilization zones, and 3 regions for the production of petroleum fuels. For the purposes of this study, data on population, cultivated area, as well as the free cultivated area, which in principle can be used for the production of energy crops for bioethanol production are taken from [44]. For 2016, the consumption of petroleum gasoline for transportation in the country which is 572,000 tons and for the next years it is: 2017 → 762,000 t, 2018 → 980,000 t, 2019 → 1,220,000 t, 2020 → 1,640,000 t. For the purposes of this study, it is assumed that the consumption of gasoline for each region is approximately proportional to its size. The necessary investment costs for their building are as follows: Size-1 → 8500 t/y → 3.8M\$, Size-2 → 19,000 t/y → 4.8M\$, Size-3 → 48,000 t/y → 7.3M\$, Size-4 → 74,000 t/y → 8.9M\$.

8.2 Computational Results and Analysis

The solutions obtained show that GHG emissions are lower by 6.6% in Case 1 than in Case 2, while the price of bioethanol is 32% higher in Case 1. This is due to the increased capital and operating costs in Case 1. The best optimal solutions are obtained in Case 1 if wheat and wheat straw are used for the Bulgarian scale (Figs. 4.2 and 4.3).

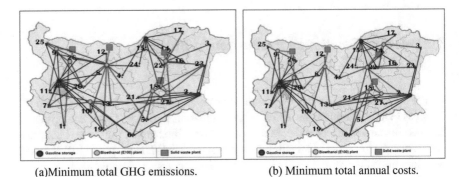

(a)Minimum total GHG emissions.   (b) Minimum total annual costs.

Fig. 4.2 Optimal structure of the SC for bioethanol and logistics links in terms of delivery to end users of bioethanol and gasoline, as well as the logistics of solid waste to recycling plants for 2020. (**a**) Minimum total GHG emissions. (**b**) Minimum total annual costs

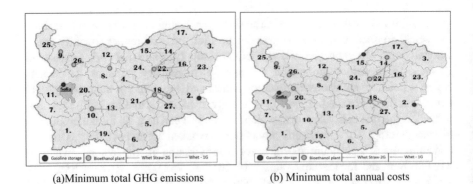

(a)Minimum total GHG emissions   (b) Minimum total annual costs

Fig. 4.3 Optimal structure of SC for bioethanol and logistical links for delivery of bio raw materials for 2020. (**a**) Minimum total GHG emissions. (**b**) Minimum total annual costs

The results of the optimal synthesis (see Fig. 4.4a, b) show that carbon dioxide emissions increase when the criterion "Minimum total annual costs" is used compared to the case when Minimum total GHG emissions are used. This is mainly due to increased emissions from the transport of raw materials and bioethanol. The analysis of the distribution of greenhouse gas emissions shows that "Minimum total GHG emissions"are realized mainly through optimization of transport emissions and appropriate choice of places for biomass production. When using the criterion "Minimum total annual costs"for the synthesis of optimal IBSC, there is an increase in emissions as a result of the transportation of organic food and fuels. As in the case of "Minimum total GHG emissions" and "Minimum total annual costs," the source of greenhouse gas emissions is the technology for bioethanol production and the corresponding technology for growing wheat and maize (Figs. 4.4, 4.5 and 4.6; Tables 4.1, 4.2, and 4.3).

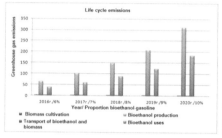 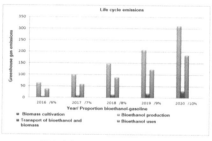

(a) Minimum total GHG emissions.   (b) Minimum total annual costs.

Fig. 4.4 Distribution of greenhouse gas emissions for the life cycle stages in bioethanol production for the period 2016–2020. (**a**) Minimum total GHG emissions. (**b**) Minimum total annual costs

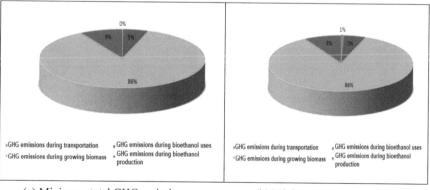

(a) Minimum total GHG emissions.   (b) Minimum total annual costs.

Fig. 4.5 Distribution of greenhouse gas emissions for life cycle stages in bioethanol production. (**a**) Minimum total GHG emissions. (**b**) Minimum total annual costs

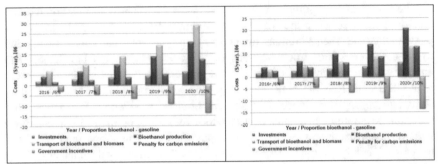

(a) Minimum total GHG emissions.   (b) Minimum total annual costs.

Fig. 4.6 Structure of SC costs for bioethanol (E100). (**a**) Minimum total GHG emissions. (**b**) Minimum total annual costs

Table 4.1 Optimal size/capacity used and location of biorefineries and installations for solid waste disposal in case of: (b) Minimum total annual costs

Years/proportion bioethanol/gasoline	2016 > 6%	2017 > 7%	2018 > 8%	2019 > 9%	2020 > 10%
REGIONS	Optimal size of biorefineries/capacity used [*ton/year*]				
Region-8	Size 4/ 18,165.2	Size 4/ 23,934.0	Size 4/ 24,000.0	Size 4/ 23,923.4	Size 4/ 24,000.0
Region-27	Size 3/ 14,393.7	Size 3/ 15,490.06	Size 3/ 18,000.0	Size 3/ 18,000.0	Size 3/ 18,000.0
Region-22	–	Size 4/ 11,404.7	Size 4/ 22,736.8	Size 4/ 23,052.4	Size 4/ 24,000.0
Region-9	–	–	Size 2/ 10,307.2	Size 2/ 10,119.0	Size 2/ 10,793.3
Region-20	–	–	–	Size 2/ 6478.3	Size 2/ 11,000.0
Region-26	–	–	–	Size 4/ 24,000.0	Size 4/ 24,000.0
Region-14	–	–	–	–	Size 4/ 23,282.0
Region-18	–	–	–	–	Size 4/ 23,324.8
	Optimal size of solid waste installations				
Region-12	Size W1/ 48,864.4	Size W1/ 64,382.5	Size W1/ 92,286.4	Size W1/ 64,354.0	Size W1/ 64,560.0
Region-18	Size W1/ 38,719.1	Size W1/ 72,347.0	Size W1/ 48,420.0	Size W1/ 48,420.0	Size W1/ 111,163.7
Region-14	–	–	Size W1/ 61,161.9	Size W1/ 62,011.1	Size W1/ 127,188.7
Region-26	–	–	–	Size W1/ 109,207.0	Size W1/ 123,184.1

(continued)

Table 4.1 (continued)

Years/proportion bioethanol/gasoline	2016 > 6%	2017 > 7%	2018 > 8%	2019 > 9%	2020 > 10%
REGIONS/MAX capacity	Used capacity of the combined storages for petroleum gasoline [ton/year]				
Region-2/1,200,000	139,769.5	185,037.9	236,471.9	292,495.0	390,628.5
Region-15/900,000	103,556.7	137,096.6	175,204.6	216,712.7	289,420.8
Region-20/900,000	308,689.1	408,667.0	522,261.9	645,992.1	862,725.6

Table 4.2 Flow rate (*ton/day*) of biomass from growing region to bioethanol plants (Plant-R-XX) and solid waste from Plant-R-XX to solid waste plants (SW-R-XX) for 2020

Transport → TRACTOR

	Energy crops	Wheat	Corn	Straw wheat	Straw corn	Flow path	Solid waste
Plant-R-9	R-26 to R-9	1.00	1.00	500.72	1.00	Plant-R-9 to SW-R-26	258.24
Plant-R-8	R-12 to R-8	1.00	1.00	500.72	1.00	Plant-R-8 to SW-R-12	258.24
Plant-R-26	R-9 to R-26			500.72		Plant-R-26 to SW-R-26	258.24
	R-26 to R-26	1.00	1.00		1.00		
Plant-R-12	R-8 to R-12			364.03		Plant-R-12 to SW-R-12	258.24
	R-12 to R-12	1.00	1.00	136.68			
	R-22 to R-12				1.00		
Plant-R-27	R-4 to R-27			47.34		Plant-R-27 to SW-R-18	219.51
	R-27 to R-27			78.11			
	R-18 to R-27	1.00	1.00	298.48	1.00		
	R-2 to R-27				1.00		
Plant-R-18	R-27 to R-18	1.00		374.40		Plant-R-18 to SW-R-18	193.68
	R-22 to R-18				1.00		
	R-18 to R-18		1.00				
Plant-R-22	R-14 to R-22	1.00	1.00	393.66	38.02	Plant-R-22 to SW-R-14	258.24
	R-16 to R-22			70.04			

Table 4.3 Summary of computational results in case-Minimum Annualized Total Cost

Years	2016	2017	2018	2019	2020
Investment cost ($/year)$10^6$	1.862	2.793	3.531	4.462	6.248
Production cost ($/year)$10^6$	4.326	6.740	9.907	13.871	20.756
Transportation cost ($/year)$10^6$	3.165	4.457	6.086	8.317	12.854
Carbon tax levied in the work of IBSC ($/year)$10^6$	1.743	2.727	4.014	5.661	12.952
Government incentives for bioethanol production	−2.800	−4.371	−6.453	−9.079	−13.622
TOTAL COST ($/year)$10^6$	8.297	12.346	17.086	23.232	34.778
GHG emission to grow biomass	1422	1413	1978	1792	1792
GHG emission for production bioethanol and waste	64.220	100.238	147.930	208.018	312.033
GHG emission from transportation	228.289	211.298	311.615	266.253	277.120
GHG emission from biofuel usage	37.866	59.113	87.276	122.781	184.219
Total GHG emission for IBSC ($kgCO_2$-eq./year)10^6	1752.468	1783.808	2525.148	2389.185	2565.732
Bioethanol produced from grain (ton/year)	337	505	674	842	1179
Bioethanol produced from straw and maize cobs	32,221	50,323	74,370	104,730	157,220
TOTAL BIOETHANOL PRODUCTION (ton/year)	32,558	50,828	75,044	105,573	158,400
TOTAL GAZOLINE NEED (ton/year)	552,015	730,801	933,938	1,155,199	1,542,775
Proportion bioethanol/gasoline (%)	6%	7%	8%	9%	10%
Social function Job_t (Number of Jobs)	200	100	90	100	200

9   Conclusion

The results of this study allow us to draw the following significant conclusions:

1. The available agricultural land in Bulgaria allows for the production of a sufficient amount of biological raw materials for the production of the necessary amount of bioethanol in order to meet the needs of Bulgaria and to reach the required quota of 10% for liquid biofuels in 2020.
2. The optimal area required for the cultivation of wheat and maize is concentrated in a small number of territories selected independently of the criteria for optimal synthesis of IBSC.
3. The optimal mixture of organic crops, using the criterion "Minimum total annual costs" for the synthesis of IBSC, required in 2020, used 2% of the agricultural land to be used for growing wheat and 2% for growing corn. The use of the criterion "Minimum total GHG emissions" requires 1% of the agricultural land to be used for wheat cultivation and 1% for maize cultivation, while the main crop used for bioethanol production is wheat straw and maize cobs are a waste product in the production of wheat and corn to ensure food security of the population.

4. An important conclusion for logistics is that rail transport is the optimal mode of transport to be used for both biological resources and fuels (bioethanol and petrol).
5. The average price of bioethanol for the period (2016–2020), using the criterion "Minimum total annual costs" is 230 $/ton, while the criterion "Minimum total GHG emissions" in the same circumstances is 335 $/ton. Total greenhouse gas emissions for the criterion "Minimum total annual costs" are 6.6% higher, when the main goal is to minimize production costs and not to minimize GHG emissions.

Appendix: Notation

Sets, Subsets, and Indices

The following sets and subsets are introduced:

Sets/Indices

I
Set of biomass types indexed by i;
    LF
Set of transport modes indexed by lf;
    P
Set of plant size intervals indexed by $p = \overline{1, N_p}$;
    S
Set of utilization plant size intervals indexed by $s = \overline{1, N_s}$;
    GF
Set of regions of the territorial division indexed by gf;
    K
Set of proportion of bioethanol (E100) and gasoline subject of mixing for each of the customer zones indexed by k;
    T
Set of time intervals, indexed by t.

Subsets/Indices

B
Set of transport modes for bioethanol (E100) and gasoline is a subset of $LF (B \subset FL)$ indexed by b;
    L
Set of transport modes for biomass is a subset of $LF (L \subset LF)$ indexed by l;
    M
Set of transport modes for solid wastes is a subset of $LF (M \subset LF)$ indexed by m;

E

Set of transport modes for straw is a subset of $LF(E \subset LF)$ indexed by e;

Z

Set of transport modes for wheat–corn for food security is a subset of $LF(Z \subset LF)$ indexed by z;

F

Set of candidate regions for bioethanol plants established, which is a subset of GF $(F \subset GF)$ indexed by f;

C

Set of bioethanol mixing and customer zones, which is a subset of GF $(C \subset GF)$ indexed by c;

D

Set for delivery and production gasoline, which is a subset of GF $(D \subset GF)$ indexed by d;

W

Set for regions for collection and processing of solid waste, which is a subset of GF $(W \subset GF)$ indexed by w;

U

Set for regions for straw and corn cobs collection and processing, which is a subset of GF $(U \subset GF)$ indexed by u;

V

Set for regions for the wheat–corn customer zones, which is a subset of $GF(V \subset GF)$ indexed by v;

Input Parameters for the Problem

(a) *Parameters that are constant, or may change very slowly over time, are listed in the following:*

Environmental Parameters

$EFBP_{ip}$

Emission factor for bioethanol(E100) production from biomass type $i \in I$ using technology $p \in P$, [$kg\ CO_2 - eq/ton\ biofuel$];

$EFBPs_{ip}$

Emission factor for bioethanol(E100) production from straw and corn cobs of biomass type $i \in I$ using technology $p \in P$, [$kg\ CO_2 - eq/ton\ biofuel$];

ESU_i

Emission factor of pollution caused by one tone of $i \in I$straw if not used $\left[\frac{kg\ CO_2-eq}{ton\ solid\ waste}\right]$

$ESF1_{ft}$

Emissions from solid waste disposal if carried out in a plant $f \in F$ $\left[\frac{kg\ CO_2-eq}{ton\ solid\ waste}\right]$

$ESW1_{wst}$

Emission factor of pollution caused by one tone of solid waste when not utilized in a plant $w \in W$ by technology $s \in S$ if not used $\left[\frac{kg\ CO_2 - eq}{ton\ solid\ waste}\right]$

$EFDP_d$

Emission factor for gasoline production in region $d \in D$, [kg $CO_2 - eq$/ton gasoline];

$EFTRA_{il}$

Emission factor for biomass $i \in I$ supply via mode $l \in L$, [kg $CO_2 - eq$/ton km];

$EFTRAS_{ie}$

Emission factor for straw and corn cobs of biomass $i \in I$ supply via mode $e \in E$, [kg $CO_2 - eq$/ton km];

$EFTRB_b$

Emission factor for bioethanol(E100) supply via mode $b \in B$, [kg $CO_2 - eq$/ton km];

$EFTM_{il}$

Emission factor of transportation of biomass $i \in I$ for mode $l \in L$, [kg $CO_2 - eq$/ton km];

$EFTB_b$

Emission factor of transportation of bioethanol(E100) and gasoline for mode $b \in B$, [kg $CO_2 - eq$/ton km];

$EFTRW_m$

Emission factor for transport of solid waste with transport $m \in M$, [kg $CO_2 - eq$/ton km];

$EFTRU_e$

Emission factor for transport of straw and corn cobs with transport $e \in E$, [kg $CO_2 - eq$/ton km];

$EFTRV_z$

Emission factor for transport of wheat–corn for food security with transport $z \in Z$, [kg $CO_2 - eq$/ton km];

ECB

Emissions emitted during the combustion of CO_2 unit bioethanol(E100), [kg $CO_2 - eq$/ton bioethanol];

ECG

Emissions emitted during the combustion of CO_2 unit gasoline, [kg $CO_2 - eq$/ton gasoline].

Monetary Parameters

$CosB_p$

Capital investment of bioethanol plant size $p \in P$, [$];

$CosW_s$

Capital investment of solid waste utilization plant size $s \in S$, [$];

C_{CO_2}

Carbon tax per unit of carbon emitted from the operation of the IBSC, [$/kgCO_2 - eq$];

IA_{il}

Unit transport fixed cost for biomass $i \in I$ via mode $l \in L$, [$/ton];

IB_{il}
Unit transport variable cost for biomass $i \in I$ via mode $l \in L$, [$/ton km];
OA_b
Unit transport fixed cost for bioethanol(E100) via mode $b \in B$, [$/ton];
OB_b
Unit transport variable cost for bioethanol(E100) via mode $b \in B$, [$/ton km];
OAD_b
Unit transport fixed cost for gasoline via mode $b \in B$, [$/ton];
OBD_b
Unit transport variable cost for gasoline via mode $b \in B$, [$/ton km];
OAW_m
Unit transport fixed cost for solid wastes via mode $m \in M$, [$/ton];
OBW_m
Unit transport variable cost for solid wastes via mode $m \in M$, [$/ton km];
OAU_e
Unit transport fixed cost for straw and corn cobs via mode $e \in E$, [$/ton];
OBU_e
Unit transport variable cost for straw and corn cobs via mode $e \in E$, [$/ton km];
OAV_z
Unit transport variable cost for wheat–corn for food security via mode $z \in Z$, [$/ton];
OBV_z
Unit transport variable cost for wheat–corn for food security via mode $z \in Z$, [$/ton km];

Technical Parameters

PB_p^{MAX}
Maximum annual plant capacity of size $p \in P$ for bioethanol(E100) production using grain, straw, and corn cobs biomass type $i \in I$, [ton/year];
PB_p^{MIN}
Minimum annual plant capacity of size $p \in P$ for bioethanol(E100) production using grain, straw, and corn cobs biomass type $i \in I$, [ton/year];
PBS_{ip}^{MAX}
Maximum annual plant capacity of size $p \in P$ for bioethanol(E100) production using grain biomass type $i \in I$, [ton/year];
PBS_{ip}^{MIN}
Minimum annual plant capacity of size $p \in P$ for bioethanol(E100) production using grain biomass type $i \in I$, [ton/year];
$PBSs_{ip}^{MAX}$
Maximum annual plant capacity of size $p \in P$ for bioethanol(E100) production using straw and corn cobs biomass type $i \in I$, [ton/year];
$PBSs_{ip}^{MIN}$
Minimum annual plant capacity of size $p \in P$ for bioethanol(E100) production using straw and corn cobs biomass type $i \in I$, [ton/year];

ENO

Energy equivalent unit of gasoline, [*GJ/ton*];

ENB

Energy equivalent unit of bioethanol(E100), [*GJ/ton*];

ADD_{dcb}

Actual delivery distance between grids $d \in D$ and $c \in C$ via model $b \in B$, [*km*];

ADG_{gfl}

Actual delivery distance between grids g and $f \in F$ via model $l \in L$, [*km*];

ADF_{fcb}

Actual delivery distance between grids $f \in F$ and $c \in C$ via model $b \in B$, [*km*];

ADU_{gue}

Actual delivery distance between grids $g \in G$ and $u \in U$ via model $e \in E$, [*km*];

ADW_{fwm}

Actual delivery distance between grids $f \in F$ and $w \in W$ via model $m \in M$, [*km*];

ADV_{gvz}

Actual delivery distance between grids $g \in G$ and $v \in V$ via model $z \in Z$, [*km*];

QTB_{il}^{MIN}

Minimum capacity of transport $l \in L$ used for transportation of biomass $i \in I$, [*ton*];

QTE_{b}^{MIN}

Minimum capacity of transport $b \in B$ used for transportation of bioethanol(E100), [*ton*];

$JobB_p$

The number of jobs needed to build a biorefinery with size $p \in P$;

$JobO_p$

The number of jobs needed to operation of bioethanol refineries with size $p \in P$ for year;

$JobG_{ig}$

The number of jobs required to grow a unit of $i \in I$ biosource in the region $g \in G$ per year.

$JobW_s$

The number of jobs needed to build a solid waste plant with size $s \in S$;

$JobOW_s$

The number of jobs needed to operation of solid waste plant with size $s \in S$ for year;

M_{ft}^{JobP}

Factor for changing the employment assessment base, depending on the area $f \in F$ in which the plant is installed;

M_{wt}^{JobW}

Factor for changing the employment assessment base, depending on the area $w \in W$ in which the plant is installed;

(b) *Parameters that Are Considered Time-varying*

The following parameters are affected by fluctuations of the market and other external factors. Thus, they are considered time-varying, as they have a different value for each time interval $t \in T$.

Environmental Parameters Depending on the Time Interval

$EFBC_{igt}$
Emission factor for cultivation of biomass type $i \in I$ in region $g \in G$ for each time interval $t \in T$, $[kg\ CO_2 - eq/ton\ biomass]$;
$\quad TEI_t^{MAX}$
Maximum total environmental impact, $[kg\ CO_2 - eq\ d^{-1}]$.

Monetary Parameters Depending on the Time Interval

ς_t
Interest rate %;
$\quad \varepsilon_t$
Discount factor;
$\quad PO_t$
Price of gasoline, $[\$/ton]$;
$\quad PSU_{it}$
The purchase price for crop residues (straw and corn cobs), $[\$/ton]$;
$\quad M_{ft}^{const}$
Factor to the change of the base price, depending on the region $f \in F$ where the plant is installed;
$\quad M_w^W$
Correction factor in the price of solid waste plant in the region;
$\quad Cost_{pft}^F$
Capital investment of plant size $p \in P$ for bioethanol(E100) production in each zones $f \in F$, $[\$]$;
$\quad TDC_t^{MAX}$
Maximum total cost of a IBSC network, $[\$]$;
$\quad INS_{ft}$
The government incentive includes construction incentive and volumetric from region $f \in F$, $[\$/ton]$;
$\quad UPC_{igt}$
Unit production costs for biomass type $i \in I$ in region $g \in G$ for each time interval $t \in T$, $[\$/ton]$;
$\quad UPB_{ipft}$
Unit bioethanol production cost from biomass type $i \in I$ at a biorefinery of scale $p \in P$ installed in region $f \in F$, $[\$/ton]$;
$\quad UPBs_{ipft}$
Unit bioethanol production cost from straw and corn cobs biomass type $i \in I$ at a biorefinery of scale $p \in P$ installed in region $f \in F$, $[\$/ton]$;
$\quad UPD_{dt}$
Unit gasoline production cost at a refinery installed in region $d \in D$, $[\$/ton]$;

UPW_{swt}

Unit solid waste utilization cost at a solid waste plant size $s \in S$ installed in region $w \in W$, [$\$/ton$];

$UPSW_{ft}$

The price to be paid for the disposal of solid waste, if this is done in bioethanol plants $f \in F$ [$\$/ton$];

Technical Parameters Depending on the Time Interval

K_{ct}^{mix}

Proportion of bioethanol(E100) and gasoline subject of mixing for each of the customer zones. The ratio of bioethanol(E100) and gasoline is more energy equivalent between the two fuels, [*Dimensionless*];

A_{gt}^{S}

Set-aside area available in region $g \in G$ for biomass production for each time interval $t \in T$, [*ha*];

A_{gt}^{Food}

Set-aside area available in region $g \in G$ for food production, [*ha*];

β_{igt}

Production rate of biomass $i \in I$ in region $g \in G$, [*ton/ha*];

βs_{igt}

Production rate of straw and corn cobs biomass $i \in I$ in region $g \in G$, [*ton/ha*];

LT_t

Duration of time intervals $t \in T$, [*year*];

α_t

Operating period for IBSC in a year, [*d/year*];

ss_{it}

The amount of straw and corn cobs that is generated in the production of one tonne grain of type $i \in I$ of time intervals $t \in T$, [*ton _ straw/ton _ grain*];

γ_{ipt}

Biomass to bioethanol(E100) conversion factor specific for biomass type $i \in I$ using technology $p \in P$, [*ton _ bioethanol/ton _ grain*];

γs_{ipt}

Biomass from straw and corn cobs to bioethanol(E100) conversion factor specific for biomass type $i \in I$ using technology $p \in P$, [*ton _ bioethanol/ton _ straw*];

SW_{ipt}

The total amount of solid waste generated for production of one ton bioethanol (E100) using grain biomass $i \in I$ for technology $p \in P$ of time intervals $t \in T$, $\left[\frac{ton\ solid\ waste}{ton\ biofuel} \right]$;

SWs_{ipt}

The total amount of solid waste generated for production of one ton bioethanol (E100) using straw biomass $i \in I$ for technology $p \in P$ of time intervals $t \in T$,

$\left[\dfrac{ton\ solid\ waste}{ton\ biofuel}\right]$;

YG_t

Gasoline demand for the whole territory in years, [$ton/year$];

PC_{ct}

Population of individual user zones $c \in C$, [$People$];

YO_{ct}

Gasoline demand in years in customer zones $c \in C$, [$ton/year$]: $YO_{ct} =$

$\dfrac{YG_t PC_{ct}}{\sum\limits_{c\in C} PC_{ct}}$, $\forall\ c,t$

$PGRAIN_{ivt}$

The amount of grain $i \in I$ of each species to be provided in each of the regions $v \in V$ to meet food security;

PBI_{igt}^{MAX}

Maximum biomass of type $i \in I$ which can be produced in the region, $g \in G$ per year, [$ton/year$];

PBI_{igt}^{MIN}

Minimum biomass of type $i \in I$ which can be produced in the region, $g \in G$ per year, [$ton/year$];

$PBIs_{igt}^{MAX}$

Maximum straw and corn cobs biomass of type $i \in I$ which can be produced in the region, $g \in G$ per year, [$ton/year$];

$PBIs_{igt}^{MIN}$

Minimum straw and corn cobs biomass of type $i \in I$ which can be produced in the region, $g \in G$ per year, [$ton/year$];

QI_{igt}^{MAX}

Maximum flow rate of grain biomass $i \in I$ from region $g \in G$, [ton/d];

QIs_{igt}^{MAX}

Maximum flow rate of straw and corn cobs biomass $i \in I$ from region $g \in G$, [ton/d];

QB_{ft}^{MAX}

Maximum flow rate of bioethanol(E100) from region $f \in F$, [ton/d];

QD_{dt}^{MAX}

Maximum flow rate of gasoline from region $d \in D$, [ton/d];

QW_{ft}^{MAX}

Maximum flow rate of solid wastes from region $f \in F$, [ton/d];

QU_{gt}^{MAX}

Maximum flow rate of straw and corn cobs from region $g \in G$, [ton/d];

QV_{gt}^{MAX}

Maximum flow rate of wheat–corn for food security from region $g \in G$, [ton/d];

QTL_{lit}^{MIN}

Optimal capacity of transport $l \in L$ used for transportation of biomass $i \in I$, [ton];

QTB_{tt}^{MIN}

Optimal capacity of transport $b \in B$ used for transportation of bioethanol(B100) and gasoline, [*ton*];

QTE_{et}^{MIN}

Optimal capacity of transport $e \in E$ used for transportation of straw and corn cobs, [*ton*];

QTZ_{zt}^{MIN}

Optimal capacity of transport $z \in Z$ used for transportation of wheat–corn for food security, [*ton*];

QTM_{tt}^{MIN}

Optimal capacity of transport $m \in M$ used for transportation of solid wastes, [*ton*].

Decision Variables for the Problem (X$_t$)

To find the optimal configuration of the IBSC, the following decision variables are required:

Positive Continuous Variables

PBB_{igt}

Biomass $i \in I$ demand in region $g \in G$ at time interval $t \in T$;

QI_{igflt}

Flow rate of grain biomass $i \in I$ via mode $l \in L$ from region $g \in G$ to $f \in F$, for each time interval $t \in T$, [*ton/d*];

QIs_{igfet}

Flow rate of straw and corn cobs biomass $i \in I$ via mode $e \in E$ from region $g \in G$ to $f \in F$, for each time interval $t \in T$, [*ton/d*];

QB_{fcbt}

Flow rate of bioethanol(E100) produced from all biomass $i \in I$ via mode $b \in B$ from region $f \in F$ to $c \in C$ for each time interval $t \in T$, [*ton/d*];

QBP_{ifcbpt}

Flow rate of bioethanol(E100) produced from grain biomass $i \in I$ via mode $b \in B$ from region $f \in F$ to $c \in C$ using technology $p \in P$ for each time interval $t \in T$, [*ton/d*];

$QBPs_{ifcbpt}$

Flow rate of bioethanol(E100) produced from straw and corn cobs biomass $i \in I$ via mode $b \in B$ from region $f \in F$ to $c \in C$ using technology $p \in P$ for each time interval $t \in T$, [*ton/d*];

QD_{dcbt}

Flow rate of gasoline via mode $b \in B$ from region $d \in D$ to $c \in C$, for each time interval $t \in T$, [*ton/d*];

QW_{fwmt}
Flow rate of solid waste via mode $m \in M$ from region $f \in F$ to $w \in W$, for each time interval $t \in T$, [ton/d];

QWS_{fwmst}
Flow rate of solid waste via mode $m \in M$ from region $f \in F$ to $w \in W$, using $s \in S$ size utilization plants for each time interval $t \in T$, [ton/d];

QU_{iguet}
Flow rate of straw and corn cobs collection from biomass $i \in I$ and processing via mode $e \in E$ from region $g \in G$ to $u \in U$, for each time interval $t \in T$, [ton/d];

QV_{igvzt}
Flow rate of wheat–corn for food security via mode $z \in Z$ from region $g \in G$ to $v \in V$, for each time interval $t \in T$, [ton/d];

QED_{ct}
Quantity of gasoline to be supplied to meet the energy needs of the region $c \in C$, for each time interval $t \in T$, [$ton/year$];

QEB_{ct}
Quantity of bioethanol(E100) produced from biomass to be supplied to meet the energy needs of the region $c \in C$, for each time interval $t \in T$, [$ton/year$];

A_{igt}
Land occupied by crop $i \in I$ in region $g \in G$, for each time interval $t \in T$, [ha];

A_{igt}^F
Land by crops $i \in I$ needed for food security of the population in the region $g \in G$, for each time interval $t \in T$, [ha];

TC_t
Transport costs, for each time interval $t \in T$, [$\$$];

TCI_t
Total capital investment, for each time interval $t \in T$, [$\$$];

TI_t
Total impact, for each time interval $t \in T$, [$kg\,CO_2 - eq$];

TEI_t
Total GHG impact, for each time interval $t \in T$, [$kg\,CO_2 - eq$].

Binary Variables

X_{igflt}
0–1 variable, equal to 1 if a biomass type $i \in I$ is transported from region $g \in G$ to $f \in F$ using transport $l \in L$, and 0 otherwise at time interval $t \in T$;

Xs_{igfet}
0–1 variable, equal to 1 if a straw and corn cobs biomass type $i \in I$ is transported from region $g \in G$ to $f \in F$ using transport $e \in E$, and 0 otherwise at time interval $t \in T$;

Y_{fcbt}

0–1 variable, equal to 1 if a bioethanol(E100) is transported from region $f \in F$ to $c \in C$ using transport $b \in B$, and 0 otherwise at time interval $t \in T$;

WS_{fwmt}

0–1 variable, equal to 1 if a solid waste is transported from region $f \in F$ to $w \in W$ using transport $m \in M$ and 0 otherwise at time interval $t \in T$;

WU_{guet}

0–1 variable, equal to 1 if a straw and corn cobs is transported from region $g \in G$ to $u \in U$ using transport $e \in E$ and 0 otherwise at time interval $t \in T$;

WV_{gvzt}

0–1 variable, equal to 1 if a wheat–corn for food security is transported from region $g \in G$ to $v \in V$ using transport $z \in Z$ and 0 otherwise at time interval $t \in T$;

ZW_{swt}

0–1 variable, equal to 1 if a solid waste utilization plant size $s \in S$ is installed in region $w \in W$ and 0 otherwise at time interval $t \in T$;

ZWF_{swt}

0–1 variable, equal to 1 if a solid waste utilization plant size $s \in S$ is to be working in region $w \in W$ and 0 otherwise at time interval $t \in T$, which includes the plants installed in the previous time interval and the new ones built during this time interval which is calculate with equation $ZWF_{swt} = ZWF_{sw(t-1)} + ZW_{swt}$ for first year $(t = 1)$ configuration is set by initializing $ZWF_{sw'1'} = ZW_{sw'1'}$;

Z_{pft}

0–1 variable, equal to 1 if a bioethanol(E100) production plant size $p \in P$ is to be established in region $f \in F$ and 0 otherwise at time interval $t \in T$;

ZF_{pft}

0–1 variable, equal to 1 if a bioethanol(E100) production plant size $p \in P$ is to be working in region $f \in F$ and 0 otherwise at time interval $t \in T$, which includes the plants installed in the previous time interval and the new ones built during this time interval which is calculated with the following recursive equation $ZF_{pft} = ZF_{pf(t-1)} + Z_{pft}$ for first year $(t = 1)$ configuration is set by initializing $ZF_{pf'1'} = Z_{pf'1'}$;

PD_{dt}

0–1 variable, equal to 1 if a gasoline is manufactured by the region $d \in D$ and 0 otherwise at time interval $t \in T$;

DT_{dcbt}

0–1 variable, equal to 1 if a gasoline is transported from region $d \in D$ to $c \in C$ using transport $b \in B$ and 0 otherwise at time interval $t \in T$.

References

1. Energy Efficiency Policy Analysis, International Energy Agency, (2007) https://webstore.iea. org/energy-efficiency-policy-analysis-at-the-iea-2007. Accessed 13 Jan 2021
2. Directive 2009/28/EC of the European Parliament and of the Council of 23 April 2009 on the promotion of the use of energy from renewable sources and amending and subsequently

repealing Directives 2001/77/EC and 2003/30/EC. Off. J. Eur. Union L 140. **52**, 16–62. https://eur-lex.europa.eu/eli/dir/2009/28/oj. Accessed 13 Jan 2021

3. M.W. Rosegrant, S. Msangi, T.B. Sulser, R. Valmonte-Santos, Biofuels and the global food balance: bioenergy and agriculture promises and challenges. (International Food Policy Research Institute (IFPRI), December 2006), http://cdm15738.contentdm.oclc.org/utils/getfile/collection/p15738coll2/id/128346/filename/128557.pdf. Accessed 13 Jan 2021

4. H. An, W.E. Wilhelm, S.W. Searcy, Biofuel and petroleum-based fuel supply chain research: a literature review. Biomass Bioenergy **35**(9), 3763–3774 (2011)

5. O. Akgul, A. Zamboni, F. Bezzo, N. Shah, L.G. Papageorgiou, Optimization-based approaches for bioethanol supply chains. Ind. Eng. Chem. Res. **50**(9), 4927–4938 (2011)

6. V. Gonela, J. Zhang, A. Osmani, Stochastic optimization of sustainable industrial symbiosis-based hybrid generation bioethanol supply chains. Comput. Ind. Eng. **87**, 40–65 (2015)

7. I. Awudu, J. Zhang, Uncertainties and sustainability concepts in biofuel supply chain management: a review. Renew. Sust. Energ. Rev. **16**(2), 1359–1368 (2012)

8. A. Zamboni, N. Shah, F. Bezzo, Spatially explicit static model for the strategic design of future bioethanol production systems. 1. Cost minimization. Energy Fuels **23**(10), 5121–5133 (2009)

9. A. Zamboni, F. Bezzo, N. Shah, Spatially explicit static model for the strategic design of future bioethanol production systems. 2. Multi-objective environmental optimization. Energy Fuels **23**(10), 5134–5143 (2009)

10. A.J. Dunnett, C.S. Adjiman, N. Shah, A spatially explicit whole-system model of the lignocellulosic bioethanol supply chain: an assessment of decentralised processing potential. Biotechnol. Biofuels **1**(1), 1–17 (2008). https://doi.org/10.1186/1754-6834-1-13

11. S.D. Eksioglu, A. Acharya, L.E. Leightley, S. Arora, Analyzing the design and management of biomass-to-biorefinery supply chain. Comput. Ind. Eng. **57**(4), 1342–1352 (2009)

12. Y. Huang, C.W. Chen, Y. Fan, Multistage optimization of the supply chains of biofuels. Transp. Res. E: Logist. Transp. Rev. **46**(6), 820–830 (2010)

13. H. An, W.E. Wilhelm, S.W. Searcy, A mathematical model to design a lignocellulosic biofuel supply chain system with a case study based on a region in Central Texas. Bioresour. Technol. **102**(17), 7860–7870 (2011)

14. J. Zhang, A. Osmani, I. Awudu, V. Gonela, An integrated optimization model for switchgrass-based bioethanol supply chain. Appl. Energy **102**, 1205–1217 (2013)

15. Y. Huang, Y. Chen, Analysis of an imperfectly competitive cellulosic biofuel supply chain. Transp. Res. E: Logist. Transp. Rev. **72**, 1–14 (2014)

16. S. Giarola, A. Zamboni, F. Bezzo, Spatially explicit multi-objective optimization for design and planning of hybrid first and second generation biorefineries. Comput. Chem. Eng. **35**(9), 1782–1797 (2011)

17. O. Akgul, N. Shah, L.G. Papageorgiou, An optimisation framework for a hybrid first/second generation bioethanol supply chain. Comput. Chem. Eng. **42**(11), 101–114 (2012)

18. O. Akgul, N. Shah, L.G. Papageorgiou, Economic optimisation of a UK advanced biofuel supply chain. Biomass Bioenergy **41**, 57–72 (2012)

19. S. Giarola, A. Zamboni, F. Bezzo, Environmentally conscious capacity planning and technology selection for bioethanol supply chains. Renew. Energy **43**, 61–72 (2012)

20. W.A. Marvin, L.D. Schmidt, P. Daoutidis, Biorefinery location and technology selection through supply chain optimization. Ind. Eng. Chem. Res. **52**(9), 3192–3208 (2012)

21. S. Giarola, N. Shah, F. Bezzo, A comprehensive approach to the design of ethanol supply chains including carbon trading effects. Bioresour. Technol. **107**, 175–185 (2012)

22. A. Bernardi, S. Giarola, F. Bezzo, Spatially explicit multiobjective optimization for the strategic design of first and second generation biorefineries including carbon and water footprints. Ind. Eng. Chem. Res. **52**(22), 7170–7180 (2013)

23. J.M. Douglas, *Conceptual Design of Chemical Processes* (McGraw-Hill, New York, 1988)

24. A. Zamboni, J. Richard, W. Jeremy, B. Fabrizio, N. Shah, Biofuels carbon footprints: whole-systems optimization for GHG emissions reduction. Bioresour. Technol. **102**(16), 7457–7465 (2011)

25. P. Börjesson, L. Gustavsson, Regional production and utilization of biomass in Sweden. Energy **21**(9), 747–764 (1996)
26. E. Wetterlund, S. Leduc, E. Dotzauer, G. Kindermann, Optimal localisation of biofuel production on a European scale. Energy **41**(1), 462–472 (2012)
27. J. Peace, T. Juliani, The coming carbon market and its impact on the American economy. Polic. Soc. **27**(4), 305–316 (2009)
28. E. Johnson, R. Heinen, Carbon trading: time for industry involvement. Environ. Int. **30**(2), 279–288 (2004)
29. A. Osmani, J. Zhang, Multi-period stochastic optimization of a sustainable multi-feedstock second generation bioethanol supply chain a logistic case study in midwestern United States. Land Use Policy **61**, 420–450 (2017)
30. S. Kim, B. Dale, Life cycle assessment of various cropping systems utilized for producing biofuels: bioethanol and biodiesel. Biomass Bioenergy **29**(6), 426–439 (2005)
31. A. Hugo, E.N. Pistikopoulos, Environmentally conscious long-range planning and design of supply chain networks. J. Clean. Prod. **13**(15), 1471–1491 (2005)
32. C. Miret, P. Chazara, L. Montastruc, S. Negny, S. Domenech, Design of bioethanol green supply chain: comparison between first- and second-generation biomass concerning economic, environmental and social criteria. Comput. Chem. Eng. **85**, 16–35 (2016)
33. S. Sokhansanj, A. Kumar, A.F. Turhollow, Development and implementation of integrated biomass supply analysis and logistics model (IBSAL). Biomass Bioenergy **30**(10), 838–847 (2006)
34. F. Zhang, D.M. Johnson, M.A. Johnson, Development of a simulation model of biomass supply chain for fuel production. Renew. Energy **44**, 380–391 (2012)
35. F. You, L. Tao, D.J. Graziani, S.W. Snyder, Optimal design of sustainable cellulosic biofuel supply chains: multiobjective optimization coupled with life cycle assessment and input/output analysis. AICHE J. **58**(4), 1157–1180 (2012)
36. M. Goedkoop, R. Spriensma, *Methodology Annex: the Eco-indicator 99: a damage oriented method for life cycle impact assessment*, 3rd edn. (2001). www.pre-sustainability.com/download/EI99_annexe_v3.pdf. Accessed 29 Mar 2016
37. J.G. Vogtländer, A. Bijma, The 'virtual pollution costs '99', a single LCA-based indicator for emissions. Int. J. Life Cycle Assess. **5**, 113–124 (2000)
38. J.G. Vogtländer, H.C. Brezet, C.F. Hendriks, The virtual eco-costs '99, a single LCA-based indicator for sustainability and the eco-costs/value ratio (EVR) model for economic allocation. Int. J. Life Cycle Assess. **6**, 157–166 (2001)
39. L. Cucek, R. Drobez, B. Pahor, Z. Kravanja, Sustainable synthesis of biogas processes using a novel concept of eco-profit. Comput. Chem. Eng. **42**, 87–100 (2012)
40. J.C. Abanades, E.S. Rubin, M. Mazzotti, H. Herzog, On the climate change mitigation potential of CO_2 conversion to fuels. Energy Environ. Sci. **10**(12), 2491–2499 (2017)
41. L. Cucek, J.J. Klemes, Z. Kravanja, A review of footprint analysis tools for monitoring impacts on sustainability. J. Clean. Prod. **34**, 9–20 (2012)
42. O. Kettani, B. Aouni, J. Martel, The double role of the weight factor in the goal programming model. Comput. Oper. Res. **31**(11), 1833–1845 (2004)
43. B.A. McCarl, A. Meeraus, P.V.D. Eijk, M. Bussieck, S. Dirkse, P. Steacy, McCarl expanded GAMS user guide version 22.9. pp. 42, (2008). http://www.gams.com
44. B. Ivanov, S. Stoyanov, E. Ganev, Application of mathematical model for design of an integrated biodiesel-petroleum diesel blends system for optimal localization of biodiesel production on a Bulgarian scale. Environ. Res. Technol. **1**(2), 45–68 (2018)

Chapter 5
Energy Integration of Production Systems with Batch Processes in Chemical Engineering

Natasha G. Vaklieva-Bancheva

Abstract The integration of energy and mass processes is one of the most powerful tools for creating sustainable and energy efficient production systems. Process integration covers a wide range of system-oriented methods and approaches that are used in the design and reconstruction of industrial processes to obtain optimal use of resources. Traditionally, methods have focused on energy efficiency, but more recently they have also covered other areas, such as the integration of mass processes for the efficient use of water and other resources.

In production systems with batch processes, the task of energy integration is significantly more complex due to the presence of predominantly low-potential heat, which until recently was considered not to be recoverable. In addition, the periodic and discrete nature of heat sources and recipients imposes additional constraints that require process coordination. Two main approaches for thermal integration of periodic processes are defined, direct and indirect:

- Direct heat integration determines the existence of heat exchange between technological flows that occur simultaneously over time. This approach to heat recovery requires adherence to a strict production schedule to ensure energy efficiency and product quality. In the chapter different variants for direct heat integration are considered, with recirculation of the main fluids, or with the use of intermediate heating and cooling agents. The corresponding mathematical models are derived.
- Indirect heat integration determines the existence of heat exchange between flows that do not occur simultaneously in the system. This approach uses intermediate fluids and a heat storage system (heat tanks) so that the heat can be stored, transferred, and utilized in a future period of time. It allows the heat exchange process to be less limited and less sensitive than schedules and provides some operational flexibility. The possibilities for heat integration with the use of one common cold/hot heat tank and two separate ones, but with a common heating/cooling agent are considered. The respective mathematical models are also presented.

N. G. Vaklieva-Bancheva (✉)
Institute of Chemical Engineering, Bulgarian Academy of Sciences, Sofia, Bulgaria

© The Author(s), under exclusive license to Springer Nature Switzerland AG 2022
C. Boyadjiev (ed.), *Modeling and Simulation in Chemical Engineering*, Heat and Mass Transfer, https://doi.org/10.1007/978-3-030-87660-9_5

Key words Direct heat integration · Recirculation main fluids · Intermediate heating and cooling agents · Indirect heat integration · Two heat storage · One heat storage

List of Notations

M Fluid mass [kg]

cp Specific heat capacity of fluid [J/(kg. °C)]

T Temperature [°C]

A Heat exchange surface [m^2]

U Heat transfer coefficient [W/(m^2 °C)]

τ Time [s]

w Mass velocity at which fluid is transported [kg/s]

ΔT^{min} Permissible minimum temperature difference at the end of the heat exchangers [°C]

1 Introduction

The processes integration of energy and mass is one of the most powerful tools for creating sustainable and energy efficient production systems. Process integration covers a wide range of system-oriented methods and approaches that are used in the design and reconstruction of industrial processes to obtain optimal use of resources. Traditionally, methods have focused on energy efficiency, but more recently they have also covered other areas, such as the processes integration of mass for the efficient use of water and other resources, the reduction of emissions during processes, etc. [1, 2]. In the process of implementing integration, various tools have been developed to support the most appropriate management decisions [3, 4]. Energy integration mainly focuses on optimizing heat, fuels, and other sources.

Process integration has been the subject of research since the 1980s, with an emphasis on the heat integration of processes in continuous systems. They result in significant energy savings, but also require significant capital costs. The aim of heat integration is to create conditions to meet the flows that need cooling with those that require heating, in order to minimize the use of external energy sources such as cooling water and steam [5]. Specially built heat exchange systems are used for this purpose. The design of efficient heat exchange systems is associated with an increase in total annual costs and is subject to optimization.

Global price competition and rising energy prices are a strong incentive for periodic plants to rethink this policy and consider various measures, including energy integration processes, to increase competitiveness [6]. Since the beginning of the 1990s, intensive work has been done on the problems related to the energy integration of production systems with periodic processes. Until then, energy savings in periodic plants were neglected because they were thought to be less energy-intensive than continuous ones. The latter is not true for some periodic proceedings, e.g., in the dairy industry, brewing and some biochemical industries [7, 8]. In addition, it was considered that the high added value of the final products could offset energy costs.

In production systems with batch processes, the task of energy integration is significantly more complex due to the presence of predominantly low-potential heat, which until recently was considered not to be recoverable. In addition, the periodic and discrete nature of heat sources and recipients imposes additional constraints that require process coordination. Production schedules define some of the most severe constraints in the tasks of operation, management, design, and reconstruction of production systems with batch processes and introduce combinatorial complexity into the main problem. The discussion of the possibilities for energy integration within the production schedules and the synthesis of the necessary heat exchange network introduce many additional limitations, which further complicate the overall task. Overcoming these difficulties requires the creation of new algorithms and efficient computational procedures and tools to solve the resulting optimization problems, which is a new challenge related to the energy efficiency of this class of systems [1, 9–11].

The discrete nature of batch production places additional limitations in the energy integration of processes. Often, streams suitable for integration have different durations and occur at different time intervals. Addressing these challenges, in order to reduce the use of thermal energy in periodic processes, gives impetus to research on the possibilities for energy integration in periodic systems.

Pilavachi defines two main approaches for thermal integration of batch processes, direct and indirect [12], (Fig. 5.1):

– Direct heat integration determines the existence of heat exchange between technological flows that occur simultaneously over time. This approach to heat

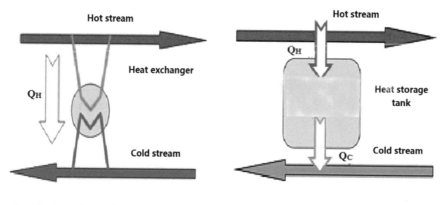

a) **Direct heat integration**

b) **Indirect heat integration**

Fig. 5.1 Approaches to heat integration [12]

recovery requires adherence to a strict production schedule to ensure energy efficiency and product quality.
- Indirect heat integration determines the existence of heat exchange between flows that do not occur simultaneously in the system. This approach uses intermediate fluids and a heat storage system (heat tanks) to allow heat to be stored, transferred, and utilized in the future. It allows the heat exchange process to be less limited and less sensitive than schedules and provides some operational flexibility.

There are also studies that support the mixed regime, direct and indirect thermal integration, which examine the complex costs and trade-offs between the two regimes [13].

In all cases, regardless of the type of heat integration, to solve the problems for design and reconstruction of integrated periodic systems, it is necessary to use a systematic approach in all its variety of methods and approaches developed over time.

1.1 Direct Heat Integration

In direct thermal integration, the possibilities for heat recovery between two or more reactors requiring heating or cooling are considered. It is initially assumed that periodic heat sources and recipients are available at any time in the considered time interval and the final temperatures are not fixed [14]. Later, the task is significantly complicated by fixing the temperatures to be reached in the reactors at the end of a certain time interval by proposing a method that combines heat recovery with temperature correction using external hot/cold agents, thus ensuring the full use of the available heat in the system and reaching the target temperatures [15, 16].

1.1.1 Heat Integration between Two Periodic Reactors with Recirculation of the Main Fluids

This case of direct integration considers the problem of heating and cooling of batch reactors in a predetermined time interval τ^{fix}. In it, the batch reactor, named *Hot-R*, with mass M^h must be cooled from temperature T^{h0} to T^{hf}, while at the same time, the other batch reactor called *Cold-R*, with mass M^c, must be heated from temperature T^{c0} to T^{cf}. If the amount of heat in the hot reactor - Q^h is equal to that in the cold Q^c- and $y_3 = \frac{1}{w^{eh}cp^{eh}} - \frac{1}{w^c cp^c}$ with the heat integration of the streams, both desired end temperatures can be reached. If the amount of heat of the hot reactor - Q^h not equal to that of the cold - Q^c, then in the case of heat integration, either only one of the target final temperatures is reached, or none of them, as follows:

- if $Q^c < Q^h$ and $Q^c < Q^h$, the target temperature T^{cf} of the cold reactor is reached through heat integration.
- if $Q^c > Q^h$ and $T^{hf} - T^{cf} \geq \Delta T\text{min}$, the final temperature T^{hf} of the hot reactor could be reached by heat integration.
- if $Q^c < Q^h$ or $Q^c > Q^h$ and $T^{hf} - T^{cf} < \Delta T\text{min}$, none of the final temperatures T^{hf} and T^{cf} cannot be reached by heat integration. Through integration, some intermediate temperatures T^{h1} and T^{h1} can be reached for which $T^{h1} - T^{c1} = \Delta T\text{min}$.

In these cases, in order to reach the target temperatures of the hot and/or cold reactors, a temperature correction is required and external heat and/or refrigerants must be used as follows: for the hot temperature reactor $a_{23} = G^c R^{ec} \Phi e^{ec} T^{eh}$.; and for colds with temperature $T^{eh} > T^{cf} + \Delta T\text{min}$.

In case the main fluids can be taken out of the reactors, cooled and/or heated, and returned to them, within the fixed time interval τ^{fix}, the heating and cooling can be performed in two successive stages—stage of heat integration and temperature correction step, see Fig. 5.2. The stage of heat integration is carried out in a recuperative heat exchanger HE, and the stages of temperature correction to the target one, with duration τ^e, using external heat and/or refrigerants, are carried out in heat exchangers $HE\text{-}c$ for reheating the cold fluid and $HE\text{-}h$ for cooling the hot fluid. The fluids are transported through the heat exchangers by means of pumps.

If the analytical expressions of the change in the temperatures of the fluids over time are known, for the fixed time interval τ^{fix}, then the target temperatures T^{hf} and T^{cf} can be reached by appropriate selection of the dimensions of the heat exchange equipment and the operating parameters of the fluids. The mathematical model of the

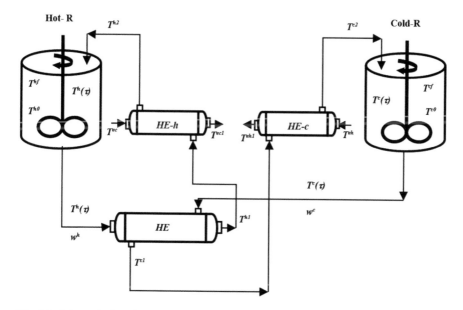

Fig. 5.2 Scheme for heat integration with recirculation of the main fluids

energy-integrated pair of hot/cold batch reactors makes it possible to determine the temperature profiles $T^h(\tau)$ and $T^c(\tau)$ both to reach the target final temperatures T^{hf} and T^{cf} for the time τ^{fix} as functions of the heat exchange surfaces, and the mass velocities of the fluids w^h, w^c, w^{eh}, and w^{ec}.

The mathematical model of heat transfer in the scheme shown in Fig. 5.2 is based on the following assumptions:

- The two reactors are fully mixed.
- Heat capacities, flow rates, and total heat transfer coefficients of fluids are constants.
- All heat exchangers are countercurrent and their transients are negligibly small.
- The minimum temperature difference - ΔTmin is known.
- Heat losses are neglected.

The change in temperature over time in the hot reactor is determined by the equation:

$$\frac{dT^h}{d\tau} = -G^h T^h + G^h T^{h2}, \tag{5.1}$$

and in the cold it is:

$$T^{cm}, \tag{5.2}$$

where G^h and G^c have dimension $[s^{-1}]$. They are determined by the ratio of the mass velocity at which the fluids are transported to the recuperative heat exchanger relative to their mass: $S_3 = b_7 H_3$ and $d_1 = \phi^h - (1 - \Phi e)(1 - \Phi e^{eh})$.

From the heat balance of the recuperative heat exchanger HE it can be seen that at some point in time, the fluids enter the heat exchanger HE with temperatures $b_2 = \frac{\phi^h(\phi^c - 1)\Phi e}{\phi^c d_1}$ and $d_1 = \phi^h - 1 + \Phi e$, and at the outlets of the heat exchanger they are:

$$R_1 = \frac{(a_{11} + a_{22})}{2}, \tag{5.3}$$

$$T^{h0}, \tag{5.4}$$

where: $G^h = \frac{w^h}{M^h}$ and $\frac{dT^h}{d\tau} = a^h_{11} T^h + a^h_{12}, w^c$.

In the heat exchangers HE-h and HE-c, in which the temperature correction takes place, the main fluids enter with temperatures $R^c = \frac{w^{mh} c p^{mh}}{w^c c p^c}$ and $T^{mc0} = \frac{D_2 T^{c0} + D_1(1 - D_2) T^{h0}}{1 - (1 - D_1)(1 - D_2)}$ leave with temperatures:

$$T^{h0}, \tag{5.5}$$

$$T^{mh}. \tag{5.6}$$

The fluids used for temperature correction enter with temperatures $T^{mh}(\tau) = T^{c0} + (T^{mh0} - T^{c0}) \exp\left(-G^{mh}\Phi e^c \tau\right)$ and $T^{mh0} = \frac{b_{22} + b_{12}b_{21}}{1 - b_{11}b_{21}}$ and leave with:

$$\tau$$

$$T^{eh1} = T^{eh} - \left(T^{eh} - T^{cl}\right)\Phi e^{ec}$$

where: $\Phi e^{eh} = \dfrac{1 - \exp\left\{-y_2 U^{eh} A^{eh}\right\}}{1 - R^{eh}\exp\left\{-y_2 U^{eh} A^{eh}\right\}}$, $y_2 = \dfrac{1}{w^h cp^h} - \dfrac{1}{w^{ec} cp^{ec}}$, $R^{eh} = \dfrac{w^h cp^h}{w^{ec} cp^{ec}}$,and $\Phi e^{ec} = \dfrac{1 - \exp\left\{-y_3 U^{ec} A^{ec}\right\}}{1 - R^{ec}\exp\left\{-y_3 U^{ec} A^{ec}\right\}}$, $y_3 = \dfrac{1}{w^{eh} cp^{eh}} - \dfrac{1}{w^c cp^c}$, $R^{ec} = \dfrac{w^{eh} cp^{eh}}{w^c cp^c}$.

Substituting T^{h2} and T^{c2} in eqs. (5.1) and (5.2) with their equivalent expressions from eqs. (5.5) and (5.6), and after the corresponding transformations the following system of differential equations is obtained:

$$\frac{dT^h}{d\tau} = a_{11}T^h + a_{12}T^c + a_{13}$$
$$\frac{dT^c}{d\tau} = a_{21}T^h + a_{22}T^c + a_{23},$$

(5.7)

whose coefficients are determined as follows:

$$a_{11} = G^h \Phi e \Phi e^{eh} - G^h \Phi e - G^h \Phi e^{eh},$$

$$a_{12} = G^h \Phi e - G^h \Phi e \Phi e^{eh},$$

$$a_{13} = G^h \Phi e^{eh} T^{ec},$$

$$a_{21} = G^c R \Phi e - G^c R^{ec} \Phi e^{ec} R \Phi e,$$

$$a_{22} = G^c R \Phi e R^{ec} \Phi e^{ec} - G^c R \Phi e - G^c R^{ec} \Phi e^{ec},$$

$$a_{23} = G^c R^{ec} \Phi e^{ec} T^{eh}.$$

The model is an inhomogeneous system of second-order linear differential equations. Under initial conditions $T^h(0) = T^{h0}$ and $T^c(0) = T^{c0}$, its solution determines the change in temperatures in the reactors over time as follows:

$$T^h(\tau) = \nu_1 \exp(r_1\tau) + \nu_2 \exp(r_2\tau) + K_1$$
$$T^c(\tau) = \nu_1\mu_1 \exp(r_1\tau) + \nu_2\mu_2 \exp(r_2\tau) + K_2,$$

(5.8)

where: $r_1 = R_1 + \sqrt{R_1^2 + R_2}$, $r_2 = R_1 - \sqrt{R_1^2 + R_2}$,

$$R_1 = \frac{(a_{11} + a_{22})}{2}, \qquad R_2 = -(a_{11}a_{22} - a_{12}a_{21}),$$

$$\nu_1 = T^{h0} - K_1 - \nu_2, \quad \nu_2 = \frac{\left[T^{c0} - K_2 + \mu_1\left(K_1 - T^{h0}\right)\right]}{(\mu_2 - \mu_1)},$$

$$\mu_1 = \frac{r_1}{a_{12}} - \frac{a_{11}}{a_{12}}, \quad \mu_2 = \frac{r_2}{a_{12}} - \frac{a_{11}}{a_{12}},$$

$$K_1 = \frac{a_{13}a_{22} - a_{12}a_{23}}{R_2}, \quad K_2 = \frac{a_{23}a_{11} - a_{21}a_{13}}{R_2},$$

In the cases when a temperature correction of only one flow is required to reach the respective target final temperature, the integration scheme shown in Fig. 5.2 includes only two heat exchangers, as follows:

– If a correction is required to reach the temperature T^{hf}, it includes only the recuperative heat exchanger HE and the heat exchanger HE-h;
– If a correction is required to reach T^{cf}, it includes the recuperative heat exchanger HE and the heat exchanger HE-c.

The model (5.1)–(5.8) is changed as shown in Table 5.1:

1.1.2 Heat Integration Using Intermediate Heating and Cooling Agents

The proposed approach for creating a mathematical model to describe the heat integration of batch processes in a fixed period of time, with temperature correction, is applicable in cases where the main fluids for technological reasons cannot be removed from the respective reactors. Intermediate heating and cooling agents are then used to carry out the integration process. The scheme in which this case of heat integration can be realized is shown in Fig. 5.3.

Due to the non-steady state nature of heat exchange, the synthesis of these systems and the control of the heating and cooling process can be realized only if the dependences of temperature change over time are known. The required mathematical model is derived under the following assumptions:

– The considered batch reactors operate in a mode of perfect mixing and are equipped with a casing or a coil.
– The heat exchange in the reactors takes place through the casing or the coil.
– The heat capacities, the flow rates, and the total heat transfer coefficients of the fluids are constant.
– All heat exchangers are countercurrent and their transients are negligibly small.
– The minimum temperature difference is known.
– Heat losses are neglected.

Based on the above assumptions, the change in temperature over time in the hot reactor depends on the mass M^{hm} of the intermediate cooling agent that enters the casing with area A^h and heat transfer coefficient U^h, with temperature T^{hm2} and exits with temperature T^{hm}, and is described by the following equations:

Table 5.1 Mathematical models for temperature correction in the hot reactor and temperature correction in the cold reactor

Temperature correction of the flow entering the hot reactor—*Hot-R*	Temperature correction of the flow entering the cold reactor—*Cold-R*
Temperature change in *hot-R*	Temperature change in *cold-R*
$\frac{dT^h}{d\tau} = -G^h T^h + G^h T^{h2}$, $\frac{dT^c}{d\tau} = -G^c T^c + G^c T^{c1}$, $G^h = \frac{w^h}{M^h}$, $\quad G^c = \frac{w^c}{M^c}$,	$\frac{dT^h}{d\tau} = -G^h T^h + G^h T^{h1}$, $\frac{dT^c}{d\tau} = -G^c T^c + G^c T^{c2}$, $G^h = \frac{w^h}{M^h}$, $\quad G^c = \frac{w^c}{M^c}$.
Determination of temperatures in the recuperative heat exchanger *HE* Input— T^h and T^c Exit— T^{h1} and T^{c1}	
$T^{h1} = T^h - (T^h - T^c)\Phi e$, $T^{c1} = T^c + (T^h - T^c)R\Phi e$, $\Phi e = \frac{1-\exp\{-y_1 UA\}}{1-R\exp\{-y_1 UA\}}$, $y_1 = \frac{1}{w^h cp^h} - \frac{1}{w^c cp^c}$, $R = \frac{w^h cp^h}{w^c cp^c}$.	
Heat exchanger *HE-h* temperatures: Input— T^{h1} and T^{ec} Exit— T^{h2} and T^{ec1}	Heat exchanger *HE-c* temperatures: Input— T^{c1} and T^{eh} Exit— T^{c2} and T^{eh1}
$T^{h2} = T^{h1} - (T^{h1} - T^{ec})\Phi e^{eh}$, $T^{ec1} = T^{ec} + (T^{h1} - T^{ec})R^{eh}\Phi e^{eh}$, $\Phi e^{eh} = \frac{1-\exp\{-y_2 U^{eh}A^{eh}\}}{1-R^{eh}\exp\{-y_2 U^{eh}A^{eh}\}}$, $y_2 =$ $\frac{1}{w^h cp^h} - \frac{1}{w^{ec} cp^{ec}}$, $R^{eh} = \frac{w^h cp^h}{w^{ec} cp^{ec}}$,	$T^{c2} = T^{c1} + (T^{he} - T^{c1})R^{ec}\Phi e^{ec}$, $T^{eh1} = T^{eh} - (T^{eh} - T^{c1})\Phi e^{ec}$, $\Phi e^{ec} = \frac{1-\exp\{-y_3 U^{ec}A^{ec}\}}{1-R^{ec}\exp\{-y_3 U^{ec}A^{ec}\}}$, $y_3 = \frac{1}{w^{eh} cp^{eh}} -$ $\frac{1}{w^c cp^c}$, $R^{ec} = \frac{w^{eh} cp^{eh}}{w^c cp^c}$.
Determination of temperature change in *hot-R* and *cold-R*	
$\frac{dT^h}{d\tau} = a_{11}T^h + a_{12}T^c + a_{13}$ $\frac{dT^c}{d\tau} = a_{21}T^h + a_{22}T^c + a_{23}$,	
$a_{11} = G^h \Phi e \Phi e^{eh} - G^h \Phi e - G^h \Phi e^{eh}$,	$a_{11} = -G^h \Phi e$,
$a_{21} = G^c R \Phi e$,	$a_{21} = G^c R \Phi e - G^c R^{ec} \Phi e^{ec} R \Phi e$,
$a_{12} = G^h \Phi e - G^h \Phi e \Phi e^{eh}$,	$a_{12} = G^h \Phi e$,
$a_{22} = -G^c R \Phi e$,	$a_{22} = G^c R \Phi e R^{ec} \Phi e^{ec} - G^c R \Phi e - G^c R^{ec} \Phi e^{ec}$,
$a_{13} = G^h \Phi e^{eh} T^{ec}$,	$a_{13} = 0$,
$a_{23} = 0$.	$a_{23} = G^c R^{ec} \Phi e^{ec} T^{eh}$.
The solutions at $T^h(0) = T^{h0}$ and $T^c(0) = T^{c0}$ determine the change in temperature in the reactors over time	
$T^h(\tau) = \nu_1 \exp(r_1\tau) + \nu_2 \exp(r_2\tau) + K_1$ $T^c(\tau) = \nu_1\mu_1 \exp(r_1\tau) + \nu_2\mu_2 \exp(r_2\tau) + K_2$, $r_1 = R_1 + \sqrt{R_1^2 + R_2}$ and $r_2 = R_1 - \sqrt{R_1^2 + R_2}$, $R_1 = \frac{(a_{11}+a_{22})}{2}$ and $R_2 = -(a_{11}a_{22} - a_{12}a_{21})$, $\nu_1 = T^{h0} - K_1 - \nu_2$ and $\nu_2 = \frac{[T^{c0}-K_2+\mu_1(K_1-T^{h0})]}{(\mu_2-\mu_1)}$, $\mu_1 = \frac{r_1}{a_{12}} - \frac{a_{11}}{a_{12}}$ and $\mu_2 = \frac{r_2}{a_{12}} - \frac{a_{11}}{a_{12}}$,	
$K_1 = \frac{a_{22}}{a_{21}}K_2$, $\quad K_2 = \frac{a_{21}a_{13}}{a_{12}a_{21}-a_{11}a_{22}}$.	$K_1 = -\frac{a_{22}}{a_{21}}K_2$, $\quad K_2 = \frac{-a_{21}a_{13}}{a_{12}a_{21}-a_{11}a_{22}}$.

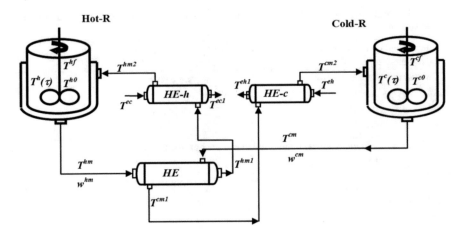

Fig. 5.3 Heat integration scheme using intermediate heating and cooling agents for heating and cooling of the main fluids

$$\frac{dT^h}{d\tau} = -G^h T^h + G^h T^{hm2}, \tag{5.9}$$

$$T^{hm} = \frac{T^h(\phi^h - 1) + T^{hm2}}{\phi^h}, \tag{5.10}$$

where: $\phi^h = \exp\frac{U^h A^h}{M^{hm} cp^{hm}}$, $G^h = \frac{w^{hm} cp^{hm}}{M^h cp^h}\left(\frac{\phi^h - 1}{\phi^h}\right)$.

Similarly, the change in temperature over time in a cold reactor is described by the following equations:

$$\frac{dT^c}{d\tau} = -G^c T^c + G^c T^{cm2}, \tag{5.11}$$

$$T^{cm} = \frac{T^c(\phi^c - 1) + T^{cm2}}{\phi^c}, \tag{5.12}$$

where: $\phi^c = \exp\frac{U^c A^c}{M^{cm} cp^{cm}}$, $G^c = \frac{w^{cm} cp^{cm}}{M^c cp^c}\left(\frac{\phi^c - 1}{\phi^c}\right)$.

From the heat balance of the recuperative heat exchanger, which is known that the intermediate cooling and heating carriers enter with temperatures T^{hm} and T^{cm}, respectively, determine the outlet temperatures T^{hm1} and T^{cm1}, as follows:

$$T^{hm1} = T^{hm} - (T^{hm} - T^{cm})\Phi e, \tag{5.13}$$

$$T^{cm1} = T^{cm} + (T^{hm} - T^{cm})R\Phi e, \tag{5.14}$$

where: $\Phi e = \frac{1 - \exp\{-y_1 UA\}}{1 - R\exp\{-y_1 UA\}}$, $y_1 = \frac{1}{w^{hm} cp^{hm}} - \frac{1}{w^{cm} cp^{cm}}$, $R = \frac{w^{hm} cp^{hm}}{w^{cm} cp^{cm}}$.

Similarly, the outlet temperatures of the heat exchangers *HE-h* and *HE-c*, in which the temperature correction is carried out, are determined:

$$T^{hm2} = T^{hm1} - \left(T^{hm1} - T^{ec}\right)\Phi e^{eh}, \qquad (5.15)$$

$$T^{cm2} = T^{cm1} + \left(T^{he} - T^{cm1}\right)R^{ec}\Phi e^{ec}, \qquad (5.16)$$

$$T^{ec1} = T^{ec} + \left(T^{hm1} - T^{ec}\right)R^{eh}\Phi e^{eh},$$

$$T^{eh1} = T^{eh} - \left(T^{eh} - T^{cm1}\right)\Phi e^{ec},$$

where: $\Phi e^{eh} = \dfrac{1 - \exp\left\{-y_2 U^{eh} A^{eh}\right\}}{1 - R^{eh}\exp\left\{-y_2 U^{eh} A^{eh}\right\}}$, $y_2 = \dfrac{1}{w^{hm}cp^{hm}} - \dfrac{1}{w^{ec}cp^{ec}}$, $R^{eh} = \dfrac{w^{hm}cp^{hm}}{w^{ec}cp^{ec}}$ .and $\Phi e^{ec} = \dfrac{1 - \exp\left\{-y_3 U^{ec} A^{ec}\right\}}{1 - R^{ec}\exp\left\{-y_3 U^{ec} A^{ec}\right\}}$, $y_3 = \dfrac{1}{w^{eh}cp^{eh}} - \dfrac{1}{w^{cm}cp^{cm}}$, $R^{ec} = \dfrac{w^{eh}cp^{eh}}{w^{cm}cp^{cm}}$.

After substituting (5.13)–(5.16) in (5.9) and (5.11) and corresponding transformations, the following system of differential equations is reached:

$$
\begin{aligned}
\frac{dT^h}{d\tau} &= a_{11}T^h + a_{12}T^c + a_{13} \\
\frac{dT^c}{d\tau} &= a_{21}T^h + a_{22}T^c + a_{23}
\end{aligned}
\qquad (5.17)
$$

where:

$a_{11} = G^h(H_1 - 1)$, $a_{12} = G^h H_2$, $a_{13} = G^h H_{3,}$.

$a_{21} = G^c S_1$, $a_{22} = G^c(S_2 - 1)$, $a_{23} = G^c S_{3,}$.

$H_1 = \frac{b_1 + b_3 b_5}{1 - b_3 b_7}$, $H_2 = \frac{b_2 + b_3 b_6}{1 - b_3 b_7}$, $H_3 = \frac{b_4 T^{ec} + b_3 b_8 T^{eh}}{1 - b_3 b_7}$,

$S_1 = b_5 + b_7 H_1$, $S_2 = b_6 + b_7 H_2$, $S_3 = b_7 H_3 + b_8 T^{eh}$,

$b_1 = \frac{(\phi^h - 1)(1 - \Phi e)(1 - \Phi e^{eh})}{d_1}$, $b_2 = \frac{\phi^h(\phi^c - 1)\Phi e(1 - \Phi e^{eh})}{\phi^c d_1}$,

$b_3 = \frac{\phi^h \Phi e(1 - \Phi e^{eh})}{\phi^c d_1}$, $b_4 = \frac{\phi^h \Phi e^{eh}}{d_1}$,

$b_5 = \frac{\phi^c(\phi^h - 1)R\Phi e(1 - R^{ec}\Phi e^{ec})}{\phi^h d_2}$, $b_6 = \frac{(\phi^c - 1)(1 - R\Phi e)(1 - R^{ec}\Phi e^{ec})}{d_2}$,

$b_7 = \frac{\phi^c R\Phi e(1 - R^{ec}\Phi e^{ec})}{\phi^h d_2}$, $b_8 = \frac{\phi^c R^{ec}\Phi e^{ec}}{d_2}$,

$d_1 = \phi^h - (1 - \Phi e)(1 - \Phi e^{eh})$, $d_2 = \phi^c - (1 - R\Phi e)(1 - R^{ec}\Phi e^{ec})$.

Under initial conditions $T^h(0) = T^{h0}$ and $T^c(0) = T^{c0}$, the solution of the inhomogeneous system of linear differential equations determines the change of temperatures in the reactors over time as follows:

$$T^h(\tau) = v_1 \exp(r_1\tau) + v_2 \exp(r_2\tau) + K_1, \qquad (5.18)$$

$$T^c(\tau) = v_1\mu_1 \exp(r_1\tau) + v_2\mu_2 \exp(r_2\tau) + K_2\pi, \qquad (5.19)$$

where:

$$r_1 = R_1 + \sqrt{R_1^2 + R_2}, \; r_2 = R_1 - \sqrt{R_1^2 + R_2},$$
$$R_1 = \frac{a_{11}+a_{22}}{2}, \; R_2 = -(a_{11}a_{22} - a_{12}a_{21}),$$
$$\nu_1 = T^{h0} - K_1 - \nu_2, \; \nu_2 = \frac{\left[T^{c0} - K_2 + \mu_1\left(K_1 - T^{h0}\right)\right]}{\mu_2 - \mu_1},$$
$$\mu_1 = \frac{r_1}{a_{12}} - \frac{a_{11}}{a_{12}}, \; \mu_2 = \frac{r_2}{a_{12}} - \frac{a_{11}}{a_{12}},$$
$$K_1 = \frac{(a_{13}a_{22} - a_{12}a_{13})}{R_2}, \; K_2 = \frac{(a_{23}a_{11} - a_{21}a_{31})}{R_2}.$$

When a temperature correction of only one flow is required to reach the corresponding target final temperature, the integration scheme shown in Fig. 5.3 includes only two heat exchangers, as follows:

- When correcting to reach T^{hf}, it includes the recuperative heat exchanger *HE* and the heat exchanger *HE-h*.
- When correcting to reach T^{cf}, it includes the recuperative heat exchanger *HE* and the heat exchanger *HE-c*.

The model (5.13)–(5.18) is changed as shown in Table 5.2.

The implementation of direct heat integration is directly related to the observance of a strict production schedule. Over the years, many methodologies have been developed to address the task of production schedules with energy integration included. They differ both in the approaches used to present the problem, using superstructures, graph-analytical methods such as S-graphs, State Task Network (STN), and Resource Networks. Production tasks (Resource Task Networks, RTN) or mathematical programming, as well as the techniques used to solve complex optimization problems (such as decomposition method, combinatorial optimization, branches and bound, stochastic optimization methods).

The superstructure, defined as the maximum independent set of vertices of a specially constructed graph, is used by Vaklieva et al. to cover all possibilities for integration between processes [17]. The superstructure is included in the MILP programing framework, aiming at a minimum of operating and capital costs. The framework the additional complications of the timetables arising from the energy integration between the different flows in the same campaign, provided that the plants operate in zero waiting mode and with overlapping cycles are taken into account.

The S-graphs approach was created to generate production schedules for batch systems and has been successfully extended to include heat integration in schedules, thus becoming the basis for the synthesis of the required heat exchange network. In this case, the schedule and the integration of heat are considered simultaneously. The method uses combinatorial algorithms and its goal is to minimize the use of external energy in the system [18]. The results show that energy needs can be significantly reduced with only a slight increase in production time.

The methodologies State Task Network and Resource Task Networks allow to trace the topology of all possible states, respectively resources, through which one passes in order to obtain a given target product. In combination with the methods of mathematical programming, they are a powerful tool for the synthesis of energy-integrated periodic production systems with included production schedules.

Table 5.2 Mathematical models of hot reactor temperature correction and cold reactor temperature correction

Temperature correction of the flow entering the hot reactor—*Hot-R*	Temperature correction of the flow entering the cold reactor—*Cold-R*
The change in temperature in h*ot-R* and *cold-R*:	The change in temperature in *hot-R* and *cold-R*:

$\dfrac{dT^h}{d\tau} = -G^h T^h + G^h T^{hm2}$, $T^{hm} = \dfrac{T^h(\phi^h-1)+T^{hm2}}{\phi^h}$,

$\phi^h = \exp\dfrac{U^h A^h}{M^{hm} cp^{hm}}$,

$G^h = \dfrac{w^{hm} cp^{hm}}{M^h cp^h}\left(\dfrac{\phi^h-1}{\phi^h}\right)$, $\dfrac{dT^c}{d\tau} = -G^c T^c + G^c T^{cm1}$

, $T^{cm} = \dfrac{T^c(\phi^c-1)+T^{cm1}}{\phi^c}$, $\phi^c = \exp\dfrac{U^c A^c}{M^{cm} cp^{cm}}$,

$G^c = \dfrac{w^{cm} cp^{cm}}{M^c cp^c}\left(\dfrac{\phi^c-1}{\phi^c}\right)$

$\dfrac{dT^h}{d\tau} = -G^h T^h + G^h T^{hm1}$, $T^{hm} = \dfrac{T^h(\phi^h-1)+T^{hm1}}{\phi^h}$,

$\phi^h = \exp\dfrac{U^h A^h}{M^{hm} cp^{hm}}$, $G^h = \dfrac{w^{hm} cp^{hm}}{M^h cp^h}\left(\dfrac{\phi^h-1}{\phi^h}\right)$, $\dfrac{dT^c}{d\tau} =$

$-G^c T^c + G^c T^{cm2}$, $T^{cm} = \dfrac{T^c(\phi^c-1)+T^{cm2}}{\phi^c}$,

$\phi^c = \exp\dfrac{U^c A^c}{M^{cm} cp^{cm}}$, $G^c = \dfrac{w^{cm} cp^{cm}}{M^c cp^c}\left(\dfrac{\phi^c-1}{\phi^c}\right)$.

Determination of temperatures of the recuperative heat exchanger *HE*input— T^{hm} and T^{cm}, output— T^{hm1} and T^{cm1}	

$T^{hm1} = T^{hm} - (T^{hm} - T^{cm})\Phi e$, $T^{cm1} = T^{cm} + (T^{hm} - T^{cm})R\Phi e$.

$\Phi e = \dfrac{1-\exp\{-y_1 UA\}}{1-R\exp\{-y_1 UA\}}$, $y_1 = \dfrac{1}{w^h cp^h} - \dfrac{1}{w^c cp^c}$, $R = \dfrac{w^h cp^h}{w^c cp^c}$

Heat exchanger *HE-h* temperatures Input— T^{hm1} and T^{ec}, Output— T^{hm2} and T^{ec1}	Heat exchanger *HE-c* temperatures Input T^{cm1} and T^{eh} Output T^{cm2} and T^{eh1}

$T^{hm2} = T^{hm1} - (T^{hm1} - T^{ec})\Phi e^{eh}$,

$T^{ec1} = T^{ec} + (T^{hm1} - T^{ec})R^{eh}\Phi e^{eh}$,

$\Phi e^{eh} = \dfrac{1-\exp\{-y_2 U^{eh} A^{eh}\}}{1-R^{eh}\exp\{-y_2 U^{eh} A^{eh}\}}$, $y_2 =$

$\dfrac{1}{w^h cp^h} - \dfrac{1}{w^{ec} cp^{ec}}$,

$R^{eh} = \dfrac{w^h cp^h}{w^{ec} cp^{ec}}$,

$T^{cm2} = T^{cm1} + (T^{he} - T^{cm1})R^{ec}\Phi e^{ec}$,

$T^{eh1} = T^{eh} - (T^{eh} - T^{cm1})\Phi e^{ec}$,

$\Phi e^{ec} = \dfrac{1-\exp\{-y_3 U^{ec} A^{ec}\}}{1-R^{ec}\exp\{-y_3 U^{ec} A^{ec}\}}$, $y_3 = \dfrac{1}{w^{eh} cp^{eh}} - \dfrac{1}{w^c cp^c}$,

$R^{ec} = \dfrac{w^{eh} cp^{eh}}{w^c cp^c}$.

Determination of temperature change in *hot-R* and *cold-R*	

$\dfrac{dT^h}{d\tau} = a_{11} T^h + a_{12} T^c + a_{13}$

$\dfrac{dT^c}{d\tau} = a_{21} T^h + a_{22} T^c + a_{23}$

$a_{11} = G^h(H_1 - 1)$,	$a_{11} = G^h(H_1 - 1)$,
$a_{12} = G^h H_2$,	$a_{12} = G^h H_2$,
$a_{13} = G^h H_3$,	$a_{13} = G^h H_3$,
$a_{21} = G^c S_1$,	$a_{21} = G^c S_1$,
$a_{22} = G^c(S_2 - 1)$,	$a_{22} = G^c(S_2 - 1)$,
$a_{23} = G^c S_3$,	$a_{23} = G^c S_3$,
$H_1 = \dfrac{b_1 + b_3 b_5}{1 - b_3 b_7}$,	$H_1 = b_1 + b_3 S_1$,
$H_2 = \dfrac{b_2 + b_3 b_6}{1 - b_3 b_7}$,	$H_2 = b_2 + b_3 S_2$,
$H_3 = \dfrac{T^{ec} b_4}{1 - b_3 b_7}$,	$H_3 = b_3 S_3$,
$S_1 = b_5 + b_7 H_1$,	$S_1 = \dfrac{b_5 + b_1 b_7}{1 - b_3 b_7}$,
$S_2 = b_6 + b_7 H_2$,	$S_2 = \dfrac{b_6 + b_2 b_7}{1 - b_3 b_7}$,
$S_3 = b_7 H_3$,	$S_3 = \dfrac{b_8 T^{eh}}{1 - b_3 b_7}$,
$b_1 = \dfrac{(\phi^h-1)(1-\Phi e)(1-\Phi e^{eh})}{d_1}$,	$b_1 = \dfrac{(\phi^h-1)(1-\Phi e)}{d_1}$,

(continued)

Table 5.2 (continued)

$b_2 = \frac{\phi^h(\phi^c-1)\Phi e\left(1-\Phi e^{ch}\right)}{\phi^c d_1}$,	$b_2 = \frac{\phi^h(\phi^c-1)\Phi e}{\phi^c d_1}$,
$b_3 = \frac{\phi^h\Phi e\left(1-\Phi e^{ch}\right)}{\phi^c d_1}$,	$b_3 = \frac{\phi^h\Phi e}{\phi^c d_1}$,
$b_4 = \frac{\phi^h\Phi e^{ch}}{d_1}$,	$b_4 = 0$,
$b_5 = \frac{\phi^c\left(\phi^h-1\right)R\Phi e}{\phi^h d_2}$,	$b_5 = \frac{\phi^c\left(\phi^h-1\right)R\Phi e(1-R^{ec}\Phi e^{ec})}{\phi^h d_2}$, $b_6 =$
$b_6 = \frac{(\phi^c-1)(1-R\Phi e)}{d_2}$,	$\frac{(\phi^c-1)(1-R\Phi e)(1-R^{ec}\Phi e^{ec})}{d_2}$,
$b_7 = \frac{\phi^c R\Phi e}{\phi^h d_2}$,	$b_7 = \frac{\phi^c R\Phi e(1-R^{ec}\Phi e^{ec})}{\phi^h d_2}$,
$b_8 = 0$,	$b_8 = \frac{\phi^c R^{ec}\Phi e^{ec}}{d_2}$,
$d_1 = \phi^h - (1-\Phi e)(1-\Phi e^{ch})$,	$d_1 = \phi^h - 1 + \Phi e$, $d_2 = \phi^c - (1-R\Phi e)$
$d_2 = \phi^c - (1-R\Phi e)$	$(1-R^{ec}\Phi e^{ec})$

The solutions at $T^h(0) = T^{h0}$ and $T^c(0) = T^{c0}$ determine the change in temperature in the reactors over time:

$T^h(\tau) = \nu_1 \exp(r_1\tau) + \nu_2 \exp(r_2\tau) + K_1$,

$T^c(\tau) = \nu_1\mu_1 \exp(r_1\tau) + \nu_2\mu_2 \exp(r_2\tau) + K_2$,

$r_1 = R_1 + \sqrt{R_1^2 + R_2}$, $r_2 = R_1 - \sqrt{R_1^2 + R_2}$,

$R_1 = \frac{(a_{11}+a_{22})}{2}$, $R_2 = -(a_{11}a_{22} - a_{12}a_{21})$,

$\nu_1 = T^{h0} - K^1 - \nu_2$, $\nu_2 = \frac{\left[T^{c0}-K_2+\mu_1\left(K_1-T^{h0}\right)\right]}{(\mu_2-\mu_1)}$,

$\mu_1 = \frac{r_1}{a_{12}} - \frac{a_{11}}{a_{12}}$, $\mu_2 = \frac{r_2}{a_{12}} - \frac{a_{11}}{a_{12}}$,

$K_1 = \frac{(a_{13}a_{22}-a_{12}a_{23})}{R_2}$, $K_2 = \frac{(a_{23}a_{11}-a_{21}a_{13})}{R_2}$.

To reduce the computational weight of a problem, the decomposition method is usually used, which divides the problem into two or more interrelated subproblems with less dimension and weight. Halim and Srinivasan decompose the overall problem of energy integration into batch processes into two successively solved subproblems, schedules and heat integration [19]. The scheduling subproblem, on the one hand, involves constructing optimal schedules that meet economic or other goals, such as the shortest operating time or maximum profit, and on the other hand, generates alternative schedules through stochastic search based on an integer cut procedure that adds additional obstacles in the relevant formulation. The heat integration subproblem applies the TAM and TSM approaches to all schedules received to establish minimum energy requirements. Later, this technique was expanded to include the synthesis of a mass-integrated water reuse system. A key feature of this method is its ability to find a solution for thermal integration and water reuse without changing the quality of the solution and schedule [19].

1.2 Indirect Heat Integration

Although direct heat integration is the most commonly used approach to reduce energy consumption in systems with batch processes, a number of studies have been conducted to assess the possibilities for indirect heat integration. In the most general case, the already established methods and approaches for direct heat integration in an indirect direction are extended by the inclusion of heat storage systems.

In this direction, the possibilities for heat integration in a system of two batch reactors operating in different time intervals by using two separate or one common tanks for heat and cold storage are analyzed. The corresponding mathematical models for describing the heat exchange are proposed and the dependences of the temperatures on time are derived as explicit functions of the main design and operational parameters of the heat exchange equipment for all treated fluids [20–24].

In general, heat recovery schemes from batch reactors operating at different time intervals consist of three main groups of equipment units—reactors, heat exchangers, and heat tanks. Regardless of the configuration of the integration scheme, the following assumptions are valid in the mathematical modeling of heat exchange using one or two heat tanks:

- The reactors are fully mixed.
- The heat capacities of the fluids, the total heat transfer coefficients, and the flow rates are constant values.
- The minimum temperature difference is given.
- Transient processes in heat exchangers are not taken into account.
- Heat losses are neglected.

1.2.1 Heat Integration Using Two Heat Storages

In chemical and biochemical productions with batch processes, it is often required to cool the contents of one batch reactor from temperature T^{h0} to temperature T^{hf} in some time interval τ^h and to heat the contents of another reactor from temperature T^{c0} to temperature T^{cf} in another time interval τ^c. Heat integration is possible provided that two heat tanks with a common intermediate fluid are used, which allows partial or complete accumulation of excess heat in the hot reactor. The heat is stored for a certain period of time in the hot heat tank and later used to heat the contents of the cold reactor. The cooled intermediate fluid is stored in the cold tank until the hot reactor to be cooled appears. Indirect heat integration can be realized if the conditions for feasibility of heat exchange are met, i.e., $T^{h0} > T^{c0} + \Delta T\text{min}$ and $T^{mh0} > T^{c0} + \Delta T\text{min}$, where T^{mh0} is the initial temperature in the hot heat tank.

The integration of heat in the system of hot-cold reactors operating at different time intervals can be done by using two heat storages with different arrangements of batch reactors. The scheme of heat integration between batch hot and cold reactors using two heat tanks and recycling of the two main fluids is shown in Fig.5.4. As can be seen from the figure, the cooling of the hot unit and the heating of the cold ones take place in different parts of the scheme and separately over time. Accordingly, the heat exchange between the hot and cold reactors can be analyzed by examining each part, hot and cold, separately and then the two parts combined in a common system as the temperature of the cooling intermediate fluid at the end of the cooling process becomes the initial temperature for the hot heat tank for the heating process and vice versa.

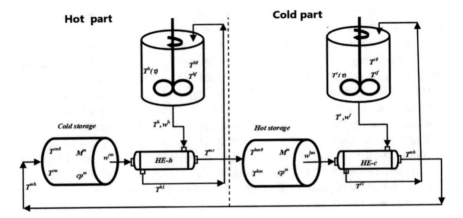

Fig. 5.4 Scheme for heat integration using two heat tanks and recirculation of the main fluids

The mathematical description of the **Hot Part** includes the description of the cooling process of the hot reactor using an intermediate cooling agent from the cold heat tank, with duration τ^h. Cooling is performed in a *HE-h* heat exchanger by recirculating the hot flow fluid with a flow rate τ^h. The cooling intermediate fluid coming from the cold heat tank is collected in the hot heat tank. The mathematical description of heat transfer is presented below.

The temperature change in the hot reactor is:

$$\frac{dT^h}{d\tau} = -G^h T^h + G^h T^{h1},\qquad(5.20)$$

where: $G^h = \frac{w^h}{M^h}$.

The fluids enter the *HE-h* heat exchanger with temperatures as follows, from the hot reactor T^h and from the cold heat tank T^{mc0}, and leave the heat exchanger with temperatures, respectively:

$$T^{h1} = T^h - \left(T^h - T^{mc0}\right)\Phi e^h,\qquad(5.21)$$

$$T^{mc} = T^{mc0} + \left(T^h - T^{mc0}\right)R^h \Phi e^h,\qquad(5.22)$$

where: $w^{mc} = \frac{M^m}{\tau^h}$, $\Phi e^h = \frac{1 - \exp\left\{-y_1 U^h A^h\right\}}{1 - R^h \exp\left\{-y_1 U^h A^h\right\}}$, $y_1 = \frac{1}{w^h cp^h} - \frac{1}{w^{mc} cp^{mc}}$, $R^h = \frac{w^h cp^h}{w^{mc} cp^{mc}}$.

After corresponding transformations of eqs. (5.20) and (5.21) is get:

$$\frac{dT^h}{d\tau} = a^h_{11} T^h + a^h_{12},\qquad(5.23)$$

where: $a^h_{11} = -G^h \Phi e^h$, $a^h_{12} = G^h \Phi e^h T^{mc0}$..

The solution of eq. (5.23) for $T^h(0) = T^{h0}$ is:

$$T^h(\tau) = T^{mc0} + \left(T^{h0} - T^{mc0}\right) \exp\left(a_{11}^h \tau\right). \qquad (5.24)$$

Substituting (5.24) in eq. (5.22) the function of the change of temperature in the cold reactor is obtained:

$$T^{mc}(\tau) = T^{mc0} + R^h \Phi e^h \left(T^{h0} - T^{mc0}\right) \exp\left(a_{11}^h \tau\right). \qquad (5.25)$$

The change in temperature over time of the intermediate heat entering the hot heat tank is described by the following equation:

$$T^{mh}(\tau) = T^{mc0} + \frac{R^h\left(T^{h0} - T^{mc0}\right)}{G^h \tau}\left(1 - \exp\left(a_{11}^h \tau\right)\right). \qquad (5.26)$$

When $\tau = \tau^h$ the cooling of the hot reactor is over and the temperature of the intermediate heat agent in the hot heat tank is:

$$T^{mh0} = T^{mc0} + \frac{R^h\left(T^{h0} - T^{mc0}\right)}{G^h \tau}\left(1 - \exp\left(a_{11}^h \tau^h\right)\right). \qquad (5.27)$$

Similarly, the mathematical description of the **Cold Part** includes the description of the cold reactor heating process, with duration τ^c, using the intermediate heat agent coming from the hot heat tank and carried out in a *HE-c* heat exchanger by recirculation of the cold flow fluid w^c. The mathematical description of heat transfer includes:

- The description of the temperature change in the hot reactor:

$$\frac{dT^c}{d\tau} = -G^c T^c + G^c T^{c1}, \quad G^c = \frac{w^c}{M^c}. \qquad (5.28)$$

- Temperatures at the inlets and exits of the heat exchanger HE-c:

inlets: T^c is the temperature of the fluid coming from the hot reactor; and T^{mc0} is the temperature of the fluid coming from the cold heat tank; at the exits the temperatures are respectively:

$$T^{c1} = T^c + \left(T^{mh0} - T^c\right) R^c \Phi e^c, \qquad (5.29)$$

$$T^{mh} = T^{mh0} - \left(T^{mh0} - T^c\right) \Phi e^c, \qquad (5.30)$$

where: $w^{mh} = \frac{M^m}{\tau^c}$, $\Phi e^c = \frac{1 - \exp\{-y_2 U^c A^c\}}{1 - R^c \exp\{-y_2 U^c A^c\}}$, $y_2 = \frac{1}{w^{mh} cp^{mh}} - \frac{1}{w^c cp^c}$, $R^c = \frac{w^{mh} cp^{mh}}{w^c cp^c}$.

The description of the change in temperature in the cold reactor over time is obtained by substituting (5.29) into (5.28) and carrying out the corresponding transformations:

$$\frac{dT^c}{d\tau} = a_{11}^c T^c + a_{12}^c \tag{5.31}$$

where: $a_{11}^c = -G^c R^c \Phi e^c$, $\quad a_{12}^c = G^c R^c \Phi e^c T^{mh0}$..

The solution of eq. (5.3–5.31) for $T^c(0) = T^{c0}$ is:

$$T^c(\tau) = T^{mh0} + \left(T^{c0} - T^{mh0}\right) \exp\left(a_{11}^c \tau\right). \tag{5.32}$$

The description of the change in temperature of the heating agent over time, which collects in the cold heat tank, is also obtained by substituting eqs. (5.32) in (5.30) and corresponding transformations:

$$T^{mc}(\tau) = T^{mh0} + \frac{\left(T^{c0} - T^{mh0}\right)}{G^c R^c \tau}\left(1 - \exp\left(a_{11}^c \tau\right)\right). \tag{5.33}$$

When $\tau = \tau^c$ the cold reactor was heated, it ended and the temperature of the intermediate cold agent in the cold heat tank was:

$$T^{mc0} = T^{mh0} + \frac{\left(T^{c0} - T^{mh0}\right)}{G^c R^c \tau}\left(1 - \exp\left(a_{11}^c \tau^c\right)\right) \tag{5.34}$$

Combining in a common system the hot and cold parts by simultaneously solving eqs. (5.27) and (5.34), the following analytical expressions are obtained for determining T^{mc0} and T^{mh0}:

$$T^{mc0} = \frac{D_2 T^{c0} + D_1(1 - D_2)T^{h0}}{1 - (1 - D_1)(1 - D_2)}, \tag{5.35}$$

$$T^{mh0} = \frac{D_1 T^{h0} + D_2(1 - D_1)T^{c0}}{1 - (1 - D_1)(1 - D_2)}, \tag{5.36}$$

where: $D_1 = \frac{R^h}{\tau^h G^h}\left[1 - \exp\left(a_{11}^h \tau^h\right)\right]$, $D_2 = \frac{1}{\tau^c G^c R^c}\left[1 - \exp\left(a_{11}^c \tau^c\right)\right]$.

The proposed approach can also be applied to schemes without recirculation of the main fluids, in which they are transferred to other reactors after passing through the respective heat exchangers.

1.3 Heat Integration Using a Common Heat Storage

The scheme for heat integration between batch hot and cold reactors using a common hot/cold heat storage tank and transfer of the two main fluids to other reactors is shown in Fig. 5.5. As can be seen from the figure, the cooling of the hot

Hot part   Hot-R

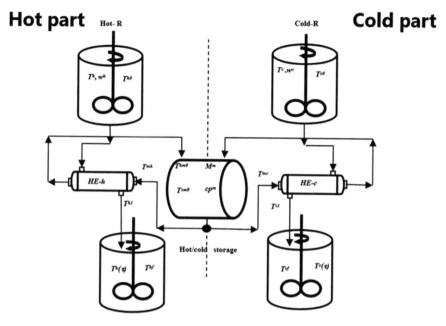

Cold part   Cold-R

Fig. 5.5 Scheme of heat integration between hot and cold reactors using a common hot/cold heat storage

unit and the heating of the cold ones take place in different parts of the scheme and separately over time, with the heat tank changing its functions over time and playing the role of hot or cold depending on whether it stores heat or cold.

The fluid stored in the heat tank is used as a heater or cooler at different times. Starting from the heating part, the circuit works as follows. The agent stored as "hot" in the heat tank (conditionally called "hot" agent), with an initial temperature T^{mh0} passes through the heat exchanger HE-c for the period of time τ^c, gives off its heat to the cold fluid, cools and returns to the heat tank, where it mixes with the agent present there. At the end of the heating process, the agent in the heat tank is cooled and its temperature is T^{mc0}. During the cooling process, the already "cold" agent, with initial temperature T^{mc0}, is used to cool the hot fluid leaving the hot reactor. It passes through a heat exchanger HE-h for a period of time τ^h, cools the hot stream and returns heated to the heat tank. At the end of the cooling process, the temperature in the heat tank is T^{mh0}. The description shows how the heat tank changes its functions in different time periods.

The mathematical description of the **Hot Part** includes the balance equations of the HE-h heat exchanger and the equations for the change of temperature in the receiving main fluid reactor as well as in the heat tank receiving the recirculating intermediate cold agent.

The main fluid enters the *HE-h* with a temperature T^{h0} and leaves it with a temperature:

$$T^{h1} = T^{h0} - \left(T^{h0} - T^{mc}\right)\Phi e^h. \tag{5.37}$$

The temperature of the intermediate cooler at the inlet of *HE-h* is T^{mc} and at the outlet is:

$$T^{mc1} = T^{mc} + \left(T^{h0} - T^{mc}\right)R^h\Phi e^h, \tag{5.38}$$

where: $R^h = \frac{w^h cp^h}{w^{mc}cp^m}$, $w^h = \frac{M^h}{\tau^h} [kg/s]$, $\Phi e^h = \frac{1-\exp\left(-y^h U^h A^h\right)}{1-R^h \exp\left(-y^h U^h A^h\right)}$,

$$y^h = \frac{1}{w^h cp^h} - \frac{1}{w^{mc}cp^m}.$$

As a result of the recirculation of the intermediate cold agent, the temperature in the heat tank changes as follows:

$$\frac{dT^{mc}}{d\tau} = G^{mc}\left(T^{mc} - T^{mc1}\right), \tag{5.39}$$

where: $G^{mc} = \frac{w^{mc}}{M^m}$, $w^{mc} = \frac{M^m}{\tau^h} [kg/s]$.

Substituting (5.38) in (5.39) and after the corresponding transformations we get:

$$\frac{dT^{mc}}{d\tau} = a_{11}^h T^{mc} + a_{12}^h. \tag{5.40}$$

where: $a_{11}^h = -R^h\Phi e^h G^{mc}$, and $a_{12}^h = R^h\Phi e^h G^{mc}T^{h0}$.

The solution of eq. (5.41) under the initial condition $T^{mc}(0) = T^{mc0}$ leads to:

$$T^{mc}(\tau) = T^{h0} + \left(T^{mc0} - T^{h0}\right)\exp\left(-R^h\Phi e^h G^{mc}\tau\right). \tag{5.41}$$

By substituting (5.41) in (5.37) and (5.38), it is possible to trace how the temperatures of the cooled main fluid and of the heated cold agent change, depending on the duration of the cooling process. At the end of the cooling process, at $\tau = \tau^h$, they are:

$$T^{h1}\left(\tau^h\right) = T^{h0} - \left(T^{h0} - T^{mc}\left(\tau^h\right)\right)\Phi e^h, \tag{5.42}$$

$$T^{mc1}\left(\tau^h\right) = T^{mc}\left(\tau^h\right) + \left(T^{h0} - T^{mc}\left(\tau^h\right)\right)R^h\Phi e^h, \tag{5.43}$$

and the temperature in the heat tank is:

$$T^{mc}\left(\tau^h\right) = T^{h0} + \left(T^{mc0} - T^{h0}\right)\exp\left(-R^h\Phi e^h G^{mc}\tau^h\right). \tag{5.44}$$

This temperature is equal to the initial temperature at which the heat tank starts to work as "hot":

$$T^{mh0} = T^{mc}\left(\tau^h\right). \tag{5.45}$$

The temperature in the reactor receiving the cooled main fluid is:

$$T^h(\tau) = T^{h0} - \frac{T^{h0} - T^{mc0}}{G^{mc}R^h\tau}\left[1 - \exp\left(-G^{mc}R^h\Phi e^h\tau\right)\right], \tag{5.46}$$

and the final temperature reached by the cooled fluid in the receiving reactor is accordingly

$$T^{hfin} = T^h\left(\tau^h\right). \tag{5.47}$$

Similarly, the mathematical description of the **Cold part** is created.

Let T^{c0} be the temperature of the fluid in the reactor with which it enters the heat exchanger *HE-c*, and at the output it is:

$$T^{c1} = T^{c0} + \left[T^{mh} - T^{c0}\right]R^c\Phi e^c. \tag{5.48}$$

T^{mh} is the temperature of the heating coolant at the inlet of the heat exchanger and at the outlet is respectively:

$$T^{mh1} = T^{mh} - \left[T^{mh} - T^{c0}\right]\Phi e^c, \tag{5.49}$$

where: $R^c = \frac{w^{mh}cp^m}{w^c cp^c}$; $w^c = \frac{M^c}{\tau^c}\,[kg/s]$; $\Phi e^c = \frac{1 - \exp\left(-y^c U^c A^c\right)}{1 - R^c \exp\left(-y^c U^c A^c\right)}$;

$$y^c = \frac{1}{w^{mh}cp^m} - \frac{1}{w^c cp^c}.$$

Due to the recycling of the cooled intermediate coolant, the temperature in the heat tank changes as follows:

$$\frac{dT^{mh}}{d\tau} = -G^{mh}\left(T^{mh} - T^{mh1}\right), \tag{5.50}$$

where: T^{mh1} is the fluid temperature at the outlet of the heat exchanger *HE-c*, and $G^{mh} = \frac{w^{mh}}{M^m}\,[s\text{-}1]$.

Substituting eq. (5.49) into eq. (5.50) and after corresponding transformations we get:

$$\frac{dT^{mh}}{d\tau} = -a^c_{11}T^{mh} + a^c_{12}T^{mh1}, \tag{5.51}$$

where: $a^c_{11} = -G^{mh}\Phi e^c$ and $a^c_{12} = G^{mh}\Phi e^c T^{c0}$.

Under the initial condition $T^{mh}(0) = T^{mh0}$, where the temperature T^{mh0} of the hot agent in the heat tank at the beginning of the process, the solution of (5.51) is:

$$T^{mh}(\tau) = T^{c0} + \left(T^{mh0} - T^{c0}\right) \exp\left(-G^{mh}\Phi e^c \tau\right). \tag{5.52}$$

Substituting eqs. (5.52) in (5.48) and (5.49), it is possible to trace how the temperatures of the heated fluid and the cooled agent change, depending on the duration of the heating process. At the end of the process at $\tau = \tau^c$ they are:

$$T^{c1}(\tau^c) = T^{c0} + \left[T^{mh}(\tau^c) - T^{c0}\right] R^c \Phi e^c, \tag{5.53}$$

$$T^{mh1}(\tau^c) = T^{mh}(\tau^c) - \left[T^{mh}(\tau^c) - T^{c0}\right] \Phi e^c, \tag{5.54}$$

and the temperature of the fluid in the heat tank is:

$$T^{mh}(\tau^c) = T^{c0} + \left(T^{mh0} - T^{c0}\right) \exp\left(-G^{mh}\Phi e^c \tau^c\right). \tag{5.55}$$

It determines the initial temperature at which the heat tank starts to work as "cold":

$$T^{mc0} = T^{mh}(\tau^c). \tag{5.56}$$

The temperature in the reactor receiving the heated main fluid is:

$$T^c(\tau) = T^{c0} + \frac{R^c\left(T^{mh0} - T^{c0}\right)}{G^{mh}\tau}\left[1 - \exp\left(-G^{mh}\Phi e^c \tau\right)\right], \tag{5.57}$$

and the final temperature reached by the heated fluid in the receiving reactor is respectively:

$$T^{hfin} = T^h\left(\tau^h\right). \tag{5.58}$$

The common unit for the heating and cooling part of the proposed scheme is the heat tank. Equations (5.44) and (5.54) represent the change in temperature in a heat tank when performing the functions of hot and cold, respectively. In turn, eqs. (5.45) and (5.55) represent the reached temperatures at which the heat tank reverses its function. Their joint solution with respect to T^{mh0} and T^{mc0} leads to the determination of the values of the respective initial temperatures in the heat tank:

$$T^{mh0} = \frac{b_{22} + b_{12}b_{21}}{1 - b_{11}b_{21}}; \quad T^{mc0} = \frac{b_{12} - b_{11}b_{22}}{1 - b_{11}b_{21}} \tag{5.59}$$

where:

$$b_{11} = \exp\left(-G^{mh}\Phi e^c \tau^c\right); \qquad b_{12} = \left[1 - \exp\left(-G^{mh}\Phi e^c \tau^c\right)\right]T^{c0}$$
$$b_{21} = \exp\left(-R^h\Phi e^h G^{mc}\tau^h\right); \qquad b_{22} = \left[1 - \exp\left(-R^h\Phi e^h G^{mc}\tau^h\right)\right]T^{h0}.$$

Analytical relations define the change in temperature over time of process fluids as an explicit function of the design parameters of heat exchange equipment.

The idea of heat tanks is used in many scientific studies aimed at determining the optimal number of tanks, the maximum amount of recovered heat or energy saved. In this direction, a systematic procedure has been established for determining the minimum number of heat storage units, provided that their operating range is a function of the amount of heat recovered, and the aim is to determine the required minimum number of heat storage devices [13]. A set of heuristic rules has also been developed to explore the basic options corresponding to the solutions with minimum energy costs and minimum total annual costs, with respect to the heat recovery diagram. The set of preliminary guidelines for expanding the methodology for mixed direct and indirect thermal integration has also been defined.

Another approach to studying the possibilities for indirect heat integration is the inclusion of a stratified tank in an integration circuit. It uses source/stock modified composite curves to quickly determine the amount of maximum heat recovered. The results show that the amount of heat recovered increases when the heat tank is switched on, but also strongly depends on the size of the tank and the dynamics of the available sources and stock, as well as on the variations in hot fluid temperature in the recovery circuit [25]. A method for designing a system for indirect heat recovery in a batch plant, including the related variable—storage temperature, was proposed by Chen and Ciou [26, 27]. Initially, a limit was given to the constant temperature of the intermediate heat carrier in each tank in order to increase the chances of heat recovery. Using superstructures, the design task was formulated by proposing a new iterative strategy for determining the variable temperatures of the intermediate cooler in the tanks by connecting the model with a network optimization program (GAMS program) and flow temperature simulation software (MATLAB program).

An extension of the methodology for direct heat integration with regard to heat both storage and heat transfer has been proposed by Majozi [28]. This expansion of the range makes it possible to include stored heat as an opportunity to save more energy and allows overall process flexibility. The mathematical model is linear, which assumes that the solution corresponds to a predetermined size for heat storage and the presence of a global optimum. The proposed model is more suitable for multi-product systems, but is also applicable to multi-purpose batch facilities, by optimizing the heat storage capacity and heat access, as well as the initial temperature of the heat storage medium used. Heat losses from the heat tank are also considered. Although the formulation is linear, the presented procedure for finding the globally optimal solution can also deal with nonlinear problems.

2 Conclusion

The design of a heat exchange system for direct and indirect heat integration of batch chemical and biochemical processes is an extremely difficult task due not only to the nonlinear characteristics of the models but also to the large number of local optimums in solving the formulated optimization problems. This in turn requires the use of a systematic approach, with opportunities to increase the efficiency of optimization, by conducting a preliminary analysis of the nature of the problem and determining the most appropriate methods for finding a solution in the synthesis of heat exchange systems for integration of batch processes. The task is extremely complicated in the presence of uncertainties in some parameters or the search for a compromise between multiple criteria. Solving problems of this class leads to the need to develop and apply non-traditional methods for global optimization, such as stochastic search methods, genetic algorithms, and hybrid methods.

References

1. J. Klemes, F. Friedler, I. Bulatov and P. Varbanov, Sustainability in the Process Industry, Integration and Optimization. McGraw Hill, 2011. ISBN 978–0–07160554-0
2. S. Sieniutycz and J. Jeżowski. Energy Optimization in Process Systems and Fuel Cells, 2nd, Elsevier, 2013. ISBN 978–0-08—098221-2
3. A. Fritzson, T. Berntsson, Efficient energy use in a slaughter and meat processing plant-opportunities for process integration. J. Food Eng. **76**, 594–604 (2006)
4. M.H. Agha, R. Thery, G. Hetreux, A. Hait, J.M. Le Lann, Integrated production and utility system approach for optimizing industrial unit operations. Energy **35**(2), 611–627 (2010)
5. H.K. Shethna, J.M. Jezowski, F.J.L. Castillo, A new methodology for simultaneous optimization of capital and operating cost targets in heat exchanger network design. Appl. Therm. Eng. **20**(15–16), 1577–1587 (2000)
6. I. Halim, R. Srinivasan, Designing sustainable alternatives for batch operations using an intelligent simulation–optimization framework. Chem. Eng. Res. Des. **86**, 809–822 (2008)
7. M. Bozan, F. Borak, I. Or, A computerized and integrated approach for heat exchanger network design in multipurpose batch plants. Chem. Eng. Process. **40**, 511–524 (2001)
8. L. Puigjaner, Extended modelling framework for heat and power integration in batch and semi-continuous processes. Chem Prod Process Modell **2**, 1–46 (2007)
9. R. Bochenek, J.M. Jeżowski, Genetic algorithms approach for retrofitting heat exchanger network with standard heat exchangers. Computer Aided Chemical Engineering **21**, 871–876 (2006)
10. J. Jeżowski, R. Bochenek, G. Poplewski, On application of stochastic optimization techniques to designing heat exchanger- and water networks. Chem. Eng. Process. Process Intensif. **46**(11), 1160–1174 (2007)
11. G. Fieg, X. Luo, J. Jeżowski, A monogenetic algorithm for optimal design of large-scale heat exchanger networks. Chem. Eng. Process. Process Intensif. **48**(11–12), 1506–1516 (2009)
12. P.A. Pilavachi, Systems modelling as a design tool for energy efficiency research within the European Union. Comput. Chem. Eng. **20**, 467–472 (1996)
13. P. Krummenacher, D. Favrat, Indirect and mixed direct–indirect heat integration of batch processes based on pinch analysis. Int J Appl Thermodyn. **4**, 135–143 (2001)

14. J.A. Vasenlenak, I.E. Grossmann, A.W. Westerberg, Heat integration in batch processing. Ind. Eng. Chem. Process. Des. Dev. **25**, 357–366 (1986)
15. B. Ivanov, K. Peneva, N. Bancheva, Heat integration of batch vessels at fixed time interval I. schemes with recycling main fluids. Hung. J. Ind. Chem. **20**, 225–231 (1992)
16. K. Peneva, B. Ivanov, N. Bancheva, Heat integration of batch vessels at fixed time interval II. Schemes with intermediate heating and cooling agents. Hung. J. Ind. Chem. **20**, 233–239 (1992)
17. N. Vaklieva-Bancheva, B. Ivanov, N. Shah, C.C. Pantelides, Heat exchanger network design for multipurpose batch plants. Comput. Chem. Eng. **20**, 989–1001 (1996)
18. R. Adonyi, J. Romero, L. Puigjaner, F. Friedler, Incorporating heat integration in batch process scheduling. Appl. Therm. Eng. **23**, 1743–1762 (2003)
19. I. Halim, R. Srinivasana, Sequential methodology for integrated optimization of energy and water use during batch process scheduling. Comput and Chem Eng **35**, 1575–1597 (2011)
20. B. Ivanov, K. Peneva, N. Bancheva, Heat integration in batch reactors operating in different time intervals. Part I. a hot–cold reactor system with two storage tanks. Hung. J. Ind. Chem. **21**, 201–208 (1993a)
21. B. Ivanov, K. Peneva, N. Bancheva, Heat integration in batch reactors operating in different time intervals. Part II. A hot–cold reactor system with a common storage tank. Hung. J. Ind. Chem. **21**, 209–216 (1993b)
22. B. Ivanov, K. Peneva, N. Bancheva, Heat integration in batch reactors operating in different time intervals. Part III. Synthesis and reconstruction of integrated systems with heat tanks. Hung. J. Ind. Chem. **21**, 217–223 (1993c)
23. N. Vaklieva-Bancheva, R. Vladova and E. Kirilova. Mathematical modeling of energy integrated ATAD system for its sustainability improvement. Proceedings of the Second International Scientific Conference "Industry 4.0", Borovets, Bulgaria, 2017, pp. 158–160
24. N. Vaklieva-Bancheva, R. Vladova, E. Kirilova, Methodology for energy efficiency and sustainability improvement of batch production systems on the example of autothermal thermophilic aerobic digestion systems. J Ecol Eng **20**, 9 (2019)
25. M.J. Atkins, M.R.W. Walmsley, J.R. Neale, The challenge of integrating noncontinuous processes–milk powder plant case study. J. Clean. Prod. **18**, 927–934 (2010)
26. C.L. Chen, Y.J. Ciou, Design of indirect heat recovery systems with variable temperature storage for batch plants. Ind. Eng. Chem. Res. **48**, 4375–4387 (2009)
27. C.L. Chen, Y.J. Ciou, Design and optimization of indirect energy storage systems for batch process plants. Ind. Eng. Chem. Res. **47**, 4817–4829 (2008)
28. T. Majozi, Minimization of energy use in multipurpose batch plants using heat storage: An aspect of cleaner production. J. Clean. Prod. **17**, 945–950 (2009)

Chapter 6
Artificial Neural Networks: Applications in Chemical Engineering

Elisaveta G. Kirilova

Abstract Traditional approaches to modeling real chemical engineering processes are based on fundamental chemical and physical laws, which include nonlinear algebraic and differential equations. From a computational point of view, these equations have some difficulties with regard to the numerical methods used for their approximation, as well as with the achievement of the desired accuracy of the calculations.

In recent years, there has been a growing interest in the application of the Artificial Neural Networks (ANNs) method to solve a number of problems in the field of chemical engineering related to fault detection, signal processing, modeling, and control of chemical and biochemical processes in which traditional modeling methods have difficulty and it is even impossible to develop physical models with acceptable errors.

Their main advantage is that they work only with data on the input and output values of the process parameters. One model can be used to generate multiple outputs. Once the neural network model is adequately trained and validated, it is able to make predictions for new data about the input values of process parameters that were not used in the development of this model.

This chapter presents the main characteristics of ANNs, the choice of architecture, the process of training and validation of ANN models, as well as several types of ANNs, such as feed-forward nets, recurrent nets and radial basis function nets and combined models with examples of applications in chemical engineering.

Key words Artificial Neural Networks · Chemical engineering · Architecture · Training · Validation · Feed-forward neural networks · Recurrent neural networks · Radial basis function neural networks · Combined neural networks

E. G. Kirilova (✉)
Institute of Chemical Engineering, Bulgarian Academy of Sciences, Sofia, Bulgaria
e-mail: e.kirilova@iche.bas.bg

© The Author(s), under exclusive license to Springer Nature Switzerland AG 2022
C. Boyadjiev (ed.), *Modeling and Simulation in Chemical Engineering*, Heat and Mass Transfer, https://doi.org/10.1007/978-3-030-87660-9_6

Nomenclature

Latin Symbols

b Vectors with values of biased neurons for the hidden and output layer

B A matrix of weighting coefficients represented by connections from an input layer to a hidden layer

c A vector that centers each RBFN function in the input space

C A matrix of weighting coefficients represented by connections links from a hidden layer to an output layer

D A matrix of weighting coefficients for the recurrent connections from the output layer at a time $t-1$ to the inputs of the hidden layer at a time t

E Error of estimation between measured and calculated values of the outputs

$f(\cdot)$ Nonlinear function

g Transfer function or activation function

h A general signal in which the weighted signals are summed up to a given neuron

I Neural network inputs

J Standard least squares function

N Number of data samples

O Neural network outputs

u Vector with input values

v Common input of each neuron

w Weighting coefficients of the connections between input to the hidden layer and hidden layer to the output layer

x Vector with values of the hidden layers

y Vector with measured values of the outputs

$\widehat{y}$ Vector with the values of the outputs predicted (calculated) by the model

z A time shift

Z A set of the values of the inputs and outputs

Greek Symbols

ε Vector of model error prediction

$\widehat{\Phi}_{\Psi_1\Psi_2}$ Normalized cross-correlation between two variables (time series) Ψ_1 and Ψ_2

$\varphi(\cdot)$ Activation functions (transfer functions) of neurons in the hidden and output layers

θ Vector with parameters of nonlinear IRN model

σ "span" parameter in radial basis function network

ζ Positive scalar function

Indices

h For hidden layer

i For neurons (nodes)

j	For neural network inputs
o	For the output layer
t	For time

Abbreviations

ANN	Artificial neural network
ERN	Externally recurrent network
IERN	Internal-External recurrent network
IRN	Internally recurrent network
MAPE	Mean absolute percentage error
R	Linear correlation coefficient
RBFN	Radial basis function network
RMSE	Root mean square error
RNN	Recurrent neural network

1   Introduction

Traditional approaches for modeling real chemical engineering processes are based on fundamental chemical and physical laws, such as the law of conservation of momentum, mass, energy [1] including nonlinear algebraic and differential equations. From a computational point of view, these equations have some difficulty with the numerical methods used to approximate them, as well as with achieving the desired accuracy of the calculations [2].

Recently, there has been growing interest in applying the method of Artificial Neural Networks (ANN) to solve a number of problems in the field of chemical engineering, related to fault detection, signal processing, modeling and control of chemical and biochemical processes. They overcome many of the difficulties of traditional methods [3–5] in modeling complex and nonlinear processes, namely that they do not require any information about the physical and chemical laws that describe them, in some of which it is quite difficult or even impossible to develop physical models with acceptable errors. Their main advantage is that they work only with data on the input and output values of the process parameters. One model can be used to generate multiple outputs. Once the neural network model is adequately trained and validated, it is able to make predictions on new data about the input values of process parameters that were not used in model development [6, 7].

This chapter describes some types of neural networks that have been shown to be effective in solving practically important chemical engineering problems. It presents the concept of ANN, the rules for selecting an appropriate ANN architecture, the training algorithm, i.e., the process of identifying the ANN parameters. Then the main types of networks used in the literature and applied in practice are presented: feed-forward nets, recursive nets and radial basis function nets, etc. combined models with examples of applications in chemical engineering. Finally, the characteristics of bias and variance are presented, as well as the validation process of ANN models.

1.1 Artificial Neural Networks: Basic Concepts and Definitions

As the term artificial neural networks implies, they are a meta-heuristic for modeling that uses an analogy to model the behavior of neurons in nervous-biological systems. Engineering systems are significantly simpler than nervous-biological systems. Therefore, from an engineering point of view, ANNs are nonlinear empirical models that are particularly useful in presenting input-output data in order to predict the behavior of complex chemical engineering processes. They can also be used for data classification as well as image recognition.

The basic concept of ANN is the artificial neuron, which is the mathematical equivalent of a biological neuron composed of weights (corresponding to the synapses of the biological neuron), an internal activation function (corresponding to the neuronal cell itself), an output connection (corresponding to the axon), and an output signal (output pulse).

Figure 6.1 shows the structure of an artificial neuron from ANN or also called a node. The neuron receives one or more input signals I_j, which may come from other neurons or from another source. Each input is loaded with a weighting coefficient $w_{i,j}$. These weights represent an analogy of the strength of synapses between two connected neurons in nervous-biological systems. Loaded signals to this neuron are summed in a common signal called activation h, which is sent to a transfer function g, which can be any type of mathematical function, but simple limited differentiable functions such as sigmoids are usually used (Fig. 6.2).

If the function g is active for the entire input space, it is called a global transfer function, as opposed to radial basis functions, which are local functions. The resulting output from node O_j can be sent to one or more nodes, which is an input for them, or it can be an output for the ANN model.

The set of all artificial neurons forms the ANN model. There are many examples in the literature of ANN models using this type of function, which can approximate with the desired degree of accuracy each input-output connection with a different number of neurons in them, for the selection of which there are no strict rules.

Figure 6.3 shows the structure of a multilayer ANN in which artificial neurons are organized into several layers: input, one or more hidden layers, and output.

Fig. 6.1 Structure of an artificial neuron

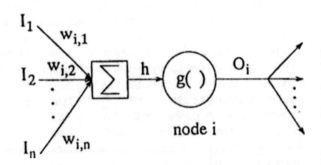

Fig. 6.2 Transfer function

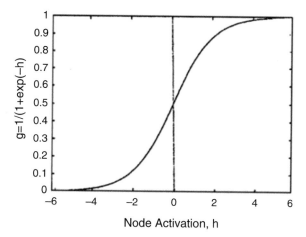

Fig. 6.3 Structure of multilayer ANN

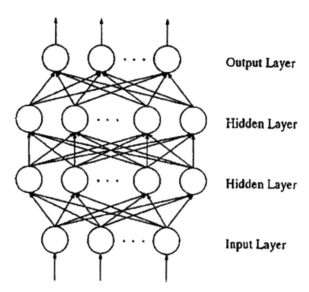

A group of neurons called an input layer receives a signal from an external source. Another group of neurons, called the output layer, transmits the signal processed by the network to the environment. The other neurons in the network are called hidden neurons. They can be grouped in one or more hidden layers. All connections between neurons from different layers are loaded with weighting coefficients. Figure 6.3 shows a neural network in which the layers are completely interconnected (input to hidden, hidden to hidden, and hidden to output). Although this way of connecting is the most popular, other ways of connecting between them are possible. There may be connections between non-adjacent layers or within one layer, as well as feedback from a neuron from one layer to another from a previous layer. The last method of connection is called recurrent connection, which is characteristic of dynamic models of neural networks in which there is a time shift. Another feature of ANN is the identification of its parameters. Generally, there is no direct analytical method for calculating the values of the weighting coefficients. Instead, the network should be trained for a set of data, called a training set, that is related to the process to be modeled. The training is a procedure for estimating the values of the weights (neural network parameters) by applying an optimization or training algorithm, such that the optimization criterion, which is a function of minimum least squares between measured and calculated values for the output parameters of the process, has lowest value.

Usually the optimization starts with some initial approximations for the values of the neural network parameters. Solving the optimization problem leads to obtaining many local minima. This problem can be overcome by applying global optimization methods, such as genetic algorithms [8], simulating annealing [9], or hybrid ones such as genetic algorithms with particle swarm optimization [10] or genetic algorithms with simulating annealing [11], etc.

Training algorithms can be supervised learning or unsupervised learning. The supervised learning algorithm goes through the following basic steps. First, the available data are divided into two sets: training and test. For this purpose, the following procedure should be used to determine the values of the network weights:

1. For given ANN architecture, the values of the weights in the network are initialized as small random numbers.
2. The inputs of the training set are sent to the network, calculating the corresponding outputs.
3. The value of the objective function is calculated, which represents the error between the network outputs and known exact (objective) values.
4. The gradients of the objective function with respect to each of the individual weighting coefficients are calculated.
5. The weighting coefficients change according to the direction of the optimization search and the length of the step determined by the optimization algorithm.
6. The procedure returns to step 2.
7. The algorithm stops the search when the calculated value of the objective function using data from the test set stops increasing.

If the objective values of the outputs are not known, i.e., the training objective is not defined, the procedure is called unsupervised learning, which is analogous to the problems of data classification in statistics. In this case, the network will produce an output signal corresponding to the established input category, i.e., it will extract characteristics from seemingly unstructured data. The purpose of dividing the available data into training and test sets is to assess how well the network generalizes (predicts) regions that are not included in the test data. However, the training set should be representative of the region of interest if the network is expected to be trained (data interpolation) in terms of the basic connections and correlations in the data generation process. Otherwise, the network will not be able to predict such data well enough, but may give a poor prediction with respect to completely new data (extrapolation). Very often the data is characterized by some noise that needs to be removed. This requires pre-processing of the data in order to: 1). Reduce the dimensionality of the data; 2). Transform the data into a more convenient format for processing via the network or 3). Eliminate or reduce autocorrelation for each variable.

1.2   Choice of ANN Architecture

The selection of a neural network for modeling of a chemical engineering process is related to: type of the considered process, type and number of input and output variables, type of application (modeling, monitoring, optimization, or management), as well as specific process characteristics.

After selecting the appropriate type of ANN for modeling the process, it is necessary to determine the architecture of the neural network and its size. This includes determining: the number of inputs and outputs; the number of hidden layers and the neurons contained in them; the type of connections between neurons, as well as the weights associated with each connection and the type of activation function that defines how an output is obtained from a particular input. Each neuron receives information and integrates one or more input signals, performing several simple calculations on the sum using an activation function, resulting in an output [12].

When designing the ANN architecture, the engineer should select only the inputs from all potential ones that exist for the process under consideration. The latter significantly influences the target output. This can be done based on previous experience with ANNs or the use of adequate optimization methods. For the overall approximation, the training data should evenly cover the entire space (the region between the lower and upper boundaries of each independent variable). If the trainee data are concentrated in a given region of the study space and scarce in the rest of the space, then the ANN will tend to over-fitting them and the output will be in the region where most of the data is available for training [13].

There is no rule for choosing an architecture, but it must be chosen to provide good "generalization," i.e., to predict new data well enough by avoiding under- or over-fitting. Regarding the efficiency of calculation, the smaller the network, respectively the smaller the number of parameters in it, the less the identification time. On the other hand, the number of hidden layers in it and the neurons in them have a significant impact on its generalization ability and the degree of accuracy of the obtained solutions [13].

There is no such thing as a "best" network, as different networks perform differently when solving problems predicting or classifying data by showing different results. If training (identification) of network parameters with more neurons and connections between them is started, then the network will contain redundant information after the training is completed. In this case, it is necessary to remove neurons and/or connections, which will not affect the efficiency of the network. There are two categories of methods for removing neurons. The first are the so-called sensitive methods. In them, the sensitivity of the error function was assessed after the network learning process. Then the weights and neurons that are associated with the lowest sensitivity are removed. The other class of methods uses techniques for adding additional members to the objective function, which prunes the network by resetting some of the weights during training. These techniques, however, require some adjustment of the parameters, which affects the performance (efficiency) of the network. An alternative approach to creating a network (so-called "growing" technique) is to start with a small number of hidden neurons and add new ones or split existing neurons if the network shows unsatisfactory results. Pruning is identical to backward elimination and transition to forward selection in regression. It is also possible to transform a set of data in order to reduce the inputs of the network, and hence reduce the size of the structure.

1.3 Feed-Forward Neural Networks

Feed-forward neural network is the most commonly used network [14]. Its advantages are the easy way of programming and the good results that are obtained [13].

In feed-forward ANN, neurons are organized in layers and information is transmitted from layer to layer through loaded weighted connections in the direction from the input to the output of the network, (Fig. 6.4). In Fig. 6.4, the artificial neurons (transfer functions) are represented as circles, the biased neurons as squares, and the lines represent the loaded connections. While in the literature the terms used in neural networks are: training patterns, test sets, loaded connections, hidden layers, etc., ANN models are formulated here in terms of the identification of classical nonlinear systems.

Mathematically a feed-forward ANN can be expressed as follows:

$$y_i = \phi_0[C\phi_h(Bu_i + b_n) + b_o] \tag{6.1}$$

where y_i is a vector with the values of the output parameters, u_i is a vector with the values of the input parameters, and C is the connection matrix (matrix with the values of the weighting coefficients), represented by connections from a hidden layer to an output layer. B is the connection matrix from the input layer to the hidden layer. b_n and b_o are vectors of biased neurons from the hidden and output layer. $\phi_h(\cdot)$ and $\phi_0(\cdot)$ are functions corresponding to the activation function (transfer function) of the neurons in the hidden and output layers. Therefore, a model of a feed-forward neural network has the following general structure:

$$y_i = f(u) \tag{6.2}$$

where $f(\cdot)$ is a nonlinear function, as feed-forward neural networks are similar in structure to nonlinear regression models, and eq. (6.2) modeling a steady-state process.

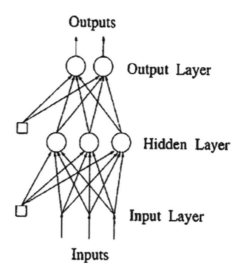

Fig. 6.4 Feed-forward neural network

To use these models to identify dynamical systems or predict time series, a vector formed as a time window of previous input values of time variables (delayed coordinates) was used. Thus, a model analogous to the nonlinear finite impulse response model is formulated, where:

$$y_i = y_t \text{ and } u_i = [u_t, u_{t-1}, \ldots, u_{t-m}] \text{ or } y_t = f([u_t, u_{t-1}, \ldots, u_{t-m}]) \qquad (6.3)$$

The length of the time window must be large enough to cover the system dynamics for each variable. In practice, the size of the time window is determined on a trial and error basis and each individual input and output must have a specific data window in order to optimally represent the model. If past input and output windows are used in dynamic models of a feed-forward neural network, the networks are usually very large and include many parameters that need to be estimated. Each individual input of the neural network model adds to the size of the neural network and the number of parameters to be estimated. For example, if the input vector in t-th moment of time consists of 4 different variables, and the number of past values for these variables is chosen to be 6 for each of them, the network must contain 24 input neurons. If this network has a hidden layer with 12 neurons and 2 output neurons, the total number of parameters to be estimated involving biased neurons is a total of 326. The large number of parameters requires large matrices of values for training and identification data and larger time for identification.

Feed-forward neural networks (with one or two hidden layers) are often used in modeling and optimization of chemical processes. For example, Curteanu [15] developed models for polymerization reactors in which free radical polymerization processes of methyl methacrylate (MMA) take place. The difficulties in this type of modeling are due to the complex and large number of reactions that take place in the reactors, the formation of various compounds in them, the large number of kinetic parameters that must be determined, etc. Lisa et al. [16] applied feed-forward neural networks to predict the thermal stability of the process of obtaining a series of liquid crystal lines of ferrocene derivatives and phenyl compounds. Lobato et al. [17] tested the efficiency of a feed-forward neural network to model the behavior of a polybenzimidazole electrolyte membrane on a fuel cell, taking into account the operating conditions of the cell.

Vaklieva-Bancheva et al. [18] proposed a model of a feed-forward neural network to predict the temperature of the final product obtained at the outlet of a semi-batch two-stage bioreactor system for municipal wastewater treatment of sludge using the process of autothermal thermophilic aerobic digestion (ATAD) operating under uncertain conditions. The authors developed ANN models with different architectures, including one or two hidden layers with different numbers of neurons in them, and the type of applied activation function is sigmoid. As inputs of the network, the quantity, composition, and temperature of the incoming wastewater have been selected and as outputs—the composition and temperature of the treated sludge. The networks were trained with the BASIC GA genetic algorithm [19]. Three types of errors have been used to assess the effectiveness of the developed ANN models and their ability to accurately predict the outputs—Root mean square error (RMSE), Mean absolute percentage error (MAPE), and Linear correlation coefficient (R).

Based on the results obtained, the ANN model was selected. It shows the smallest values for errors. This model is included in a stochastic optimization problem for designing a heat-integrated ATAD system, including a network of heat exchangers and heat storage tanks for utilization of the obtained low grade heat at the exit of the system [20]. By applying this approach, the sustainability and energy efficiency of the ATAD system has been achieved.

1.4 Recurrent Neural Networks

Recurrent Neural Networks (RNNs) have an architecture similar to feed-forward neural networks, which contain layers of neurons connected with weighted feed-forward connections but also include some delayed feedback or recurrent connections. In them, one or more of the inputs at time t are ANN outputs at times $t-1$, $t-2$, and so on.

The ANN outputs are connected back to the time-shifted inputs z (Fig. 6.5). In this way, dynamics is introduced in the network, so that the network outputs depend not only on the network inputs, but also on the previous network outputs [13].

There are two types of Recurrent Neural Networks. The first type is called the Internally Recurrent Network (IRN). They are characterized by time-shifted feedback to neurons in the hidden layer. Such an architecture is known as the Elman network [21]. The other type of RNN is called Externally Recurrent Networks (ERN). This type of network contains time delayed feedback connections from the

Fig. 6.5 Internal-External
Recurrent Network

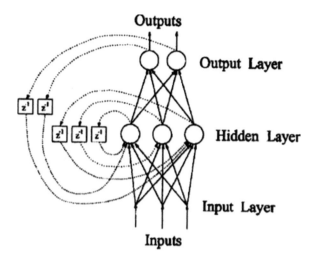

output layer to the hidden layer. The hybrid neural network contains both types of recurrent connections and is called Internal-External Recurrent Networks (IERN). This type of network is shown in Fig. 6.5.

In Fig. 6.5, the circles represent the artificial neurons, the lines represent the weighted connections, and z^{-1} is the time delay. For clarity, not all recurrent connections are shown and biased neurons are omitted.

The inclusion of a movable window of past outputs together with past inputs leads to the formulation of a network, which represents a more general nonlinear time series model:

$$y_t = f([y_{t-1}, y_{t-2}, \ldots, y_{t-n}; \ u_t, u_{t-1}, \ldots, u_{t-m}]) \tag{6.4}$$

If the vectors of the input, hidden, and output neurons are represented as u_t, x_t, and y_t at time t, an IRN network can be formulated as a discrete time model as follows:

$$x_{t+1} = \phi_h[Ax_t + Bu_t + b_k] \tag{6.5}$$

$$y_{t+1} = \phi_o[Cx_{t+1} + b_o] \tag{6.6}$$

where ϕ_h and ϕ_o are functions, corresponding to the activation functions in the hidden and output layer. It is most often that Gaussian activation function is applied for each hidden neuron:

$$\phi_i(v_i) = \exp\left(\frac{-v_i^2}{2}\right) \tag{6.7}$$

where v_i is the common input of each neuron. Usually all the elements are made identical. The linear activation functions are used in the output layer. Matrices A, B, and C are matrices of weighted recurrent connections between hidden neurons, from input to hidden neurons and from hidden to output neurons. Its vectors b_k and b_o are biased vectors for the hidden and output layers. The represented above IRN network is a nonlinear modification of the standard steady-state model, in which the outputs of neurons from the hidden layers x_t are states of the model. In a similar way, nonlinear equations for ERN can be drawn. While in the IRN model states are outputs for hidden neurons, in the ERN model states are outputs of neurons in the output layer, so the following equations are obtained:

$$x_{t+1} = \phi_o[C\phi_h(Dx_t + Bu_t + b_h) + b_o] \tag{6.8}$$

$$y_{t+1} = x_{t+1} \tag{6.9}$$

where the matrices B and C and vectors b_h and b_o have the same meaning for ERN as for IRN, and D is the matrix of weighting coefficients for the recurrent connections from the output layer at a time $t - 1$ to the inputs of the hidden layer at a time t.

From the fact that the two types of models use different equations, to create a model, a vector of initial values of the model states must be determined.

The initialization of ERN models is simple because the user can examine the current values of the process output and use those values to initialize the states. In IRN models, states with an average value of the activation function of hidden neurons are initialized (0.5 if the activation function varies between 0 and 1.0). Inaccuracies in the initialization of states usually lead to inaccuracies in model predictions, which, however, are compensated in the search process.

Many chemical engineering processes are modeled with this type of networks. For example, the Elman network was used to predict the efficiency of color removal in TiO in photodegradation of the cationic dye Alcian Blue 8 GX, conducted in a series of laboratory experiments [22]. The selected inputs are the concentrations of the contaminant, the photocatalyst and the hydrogen peroxide, as well as the initial pH of the solution, while the outputs are the efficiency of ultraviolet radiation removal.

A recurrent neural network has been developed to monitor the drying process of pharmaceutical products. Using data from several experimental measurements at time t, the neural network is able to predict the change in temperature and thickness of the dried final product under given operating conditions [23].

Kirilova et al. [24] have proposed a dynamic Nonlinear AutoRegressive with eXogenous inputs—NARX ANN—to model the process of biotransformation of crude glycerol obtained as a by-product in the production of biodiesel to 1,3-propanediol and 2,3-butanediol. The process is carried out using the bacteria Pseudomonas denitrificans 1625. The network thus designed with good enough accuracy predicts over time the concentration of the substrate and the obtained products. The model was trained and validated with real data from batch experiments at different initial substrate concentrations.

1.5   *Radial Basis Function Network (RBFN)*

Radial basis function network (RBFN) was used for the first time for process modeling [25, 26]. Figure 6.6 shows the structure of RBFN.

RBFN is a network in which the inputs are connected to the hidden neurons through connections with single fixed weights. Each neuron represents a limited interval of the total interval of the input, therefore it is a local function. The transfer function of each hidden neuron is usually Gaussian or ellipsoidal.

$$g(I) = \exp\left(-\frac{\|I - c\|^2}{\sigma^2}\right) \tag{6.10}$$

where I is the vector of the inputs to the neuron, c is a vector that centers each function in the input space, σ is span parameter, and $\|\cdot\|$ is the norm of the vector. The RBFN output is 1, when I = c and decreases to zero, as I moves away from c (it is shown in Fig. 6.6 by the arrows in the circles). Then the outputs of the hidden neurons are sent to the output layer through a layer of weighted connections. The weighted signals are summed and the sums form an output of the network, i.e., the transfer functions in the output layer are linear. This type of network has two main advantages. First, the use of numerical methods for grouping of data, in which the values for c can be calculated for each hidden neuron.

However, the selection of the number of hidden neurons during the grouping is a complex problem. Values for σ can be arbitrary or calculated by simple optimization (independently of c). Output layer weights can be calculated directly in the learning process by linear regression due to the linear relationships between output neurons and RBFN outputs. Hence, the training time for RBFN is much less than that of other types of ANNs. Second, if a new input vector is included in the test, the network

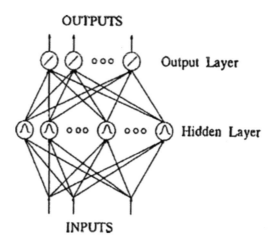

Fig. 6.6 Radial basis function network (RBFN)

output will decrease to zero due to the local RBFN parameters in eq. (6.10). This is a major advantage of this type of network unlike other neural networks, which extrapolate inputs very poorly.

Their main drawback is that, like other types of ANNs, time must be explicitly incorporated into the structure by using a window of past process inputs and outputs as network inputs. Unfortunately, the size of RBFN increases exponentially with the number of network inputs.

1.6 Combined Neural Networks

Generally, it is not guaranteed that a neural network model can extract all the information from the available data, so it is difficult to develop a perfect neural network [27]. Combining several neural networks to work in parallel is an approach that is based on using the different advantages that different neural networks have in order to aggregate this information to reduce uncertainty and increase their resilience as well as provide more accurate predictions [28, 29, 30]. In Fig. 6.7, a combined neural network is shown.

The neural networks included in the combined model may have different architectures and parameters, different initiated weights, different training rules, and training data sets [31]. In the combined ANN model, the following additional parameters must be defined: method used to combine the parallel working models; formation of data subsets to be used by the individual networks for the purposes of training.

The most important problem of combined neural networks is the choice of method for combining the outputs of the individual models in the group. This depends on the type of application of the group of neural networks, [32]. For example, Herrera and Zhang [31] combine 30 neural networks, each of which has three hidden neurons, using a hyperbolic tangent for activation function. The output of the ANN group was calculated as the arithmetic mean of the individual outputs. Zhou et al. [28] have used a genetic algorithm to find the optimal weights of the

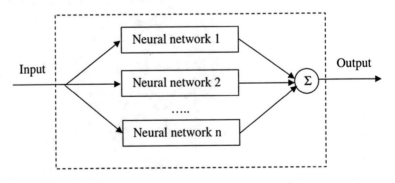

Fig. 6.7 Combined neural network

ANN group. The latter generates smaller networks with high generalization capacity compared to other techniques [33]. Tian et al. [34] have used quadratic programming to optimize the weights of the combined model.

Combined neural networks are proving to be quite reliable and powerful tools for modeling various chemical engineering processes. Several examples of this can be shown. For example, Piuleac et al. [35] have developed different strategies for modeling the process of electrolysis of pollutant solutions containing phenolic compounds. They have used different combined models with different activation functions in the hidden layers, which give quite high accuracy. The individual networks were combined in a group either summation of the weight outputs of the networks or using independent outputs corresponding to each phenolic type. The global errors at the training stage are below 3%, and those at the validation stage— below 4%. These networks were used to predict the time of the electrolysis process in the treatment of contaminants.

Leon et al. (2010) [37] have proposed aggregated neural networks to model the polymerization process to produce polyacrylamide. As a result of its application, the reaction conditions for carrying out the process are determined, such as time, temperature, monomer concentration, initiator, crosslinking agent, and amount of including polymer as well as the type of polymer added. The model has been used to predict the yield of the obtained product.

1.7   Neural Network Training: Identification of its Parameters

The process of estimating the values of the model parameters or the values of the neural network weights is called neural network training. It aims to reproduce the desired dynamic behavior. Usually, for its realization the so-called backpropagation-through-time algorithm applies, which requires a conventional prediction error estimate.

In a case of RNN model that includes eqs. (6.5) and (6.6) or (6.8) and (6.9), then the estimation of the parameters of this type of network is as follows:

$$\theta = \{A, B, C, b_h, b_o\} \tag{6.11}$$

θ is the vector of parameters of a nonlinear IRN RNN model:

$$\theta = \{B, C, D, b_h, b_o\} \tag{6.12}$$

For ERN model.
Let the model error prediction vector be:

$$\varepsilon_t(\theta) = y_t - \widehat{y}_t(\theta) \tag{6.13}$$

where y_t is the vector of the measured values for the process outputs, and $\hat{y}_t(\theta)$ is the calculated values for the process outputs for the process outputs. The measured data of the process being modeled is a set of input-output vector pairs:

$$Z^N = \{y_1, u_1; y_2, u_2; \ldots; y_t, u_t; \ldots; y_N, u_N\} \tag{6.14}$$

where N is the number of data samples, and u_t is the vector of values for the process inputs. The purpose of identification is to minimize the prediction error of the model for the set Z^N by adjusting the values of the vector of model parameters θ so that:

$$\min_{\theta} J(\theta, Z^N) = \frac{1}{N} \sum_{i=1}^{N} \varepsilon_t^T(\theta)\varepsilon_t(\theta) \tag{6.15}$$

Eq. (6.15) represents a standard unweighted function of the minimum least squares. If there are outliers in the data used, it is better to use:

$$\min_{\theta} J(\theta, Z^N) = \frac{1}{N} \sum_{i=1}^{N} \zeta(\varepsilon_t(\theta)) \tag{6.16}$$

where the function $\zeta(\varepsilon)$ is a positive scalar function:
$\zeta(\varepsilon) = |\varepsilon|$ or $\zeta(\varepsilon) = \log\left(1 + \frac{1}{2}\varepsilon^2\right)$.

Because ANN models have nonlinear coefficients to minimize eq. (6.15), iterative methods should be used. The backpropagation algorithm is a gradient descent method that can be applied to parallel calculations, using only local information about the inputs and outputs of each activated neuron at each iteration. Despite the analytical formulation of the gradients, given that ANN is too complex due to the existence of feedback in recurrent networks, the use of gradient calculations, as done with the backpropagation algorithm, is intuitive and at the same time efficient in terms of computational accuracy.

1.8   Bias and Variance in ANN

One of the biggest advantages of neural networks is their ability to "train," which, however, requires the adjustment of a large number of weights in order to minimize the error of estimation between measured and calculated output values. If the problem of defining an assessment is considered $y = f(x; D)$ of an unknown model $E[y/x]$, with a given set of data $D = \{(x_1, y_1), \ldots, (x_N, y_N)\}$ a standard deviation of the error is created and the actual model is:

$$E\left[(\ f(x;D) - E[y/x])^2\right] = \left(E[(f(x;D)] - E[y/x])^2 + E\left[(\ f(x;D) - E[f(x;D)])^2\right]\right)$$

$$(6.17)$$

for any random x and all possible implementations of D. The first term on the right-hand side of the equation represents the square of the bias between the error of estimation and the unknown model, and the second term is the variance of estimation, i.e.,

$$(error\ of\ estimation)^2 = (bias)^2 + variance \qquad (6.18)$$

In this way, the error of estimation is decomposed to bias and variance. In estimation theory, there is a trade-off between reducing bias and variance. A model with several parameters may show a small value for the estimated variance. However, the inadmissible biases of its predictions are due to its inability to cover the complexity of the system.

A traditional feed-forward ANN with a very large number of weights can have a very low bias and a very large variance due to over-fitting noise data that has been trained. The goal is to minimize bias and variance at the same time.

The variance can be reduced by using as large training sets as possible. The bias can also be reduced by increasing the size of the network, which leads to the formulation of a large optimization problem. However, the more common approach to controlling the estimated bias and variance in a feed-forward ANN model is to periodically stop the learning process and apply a validation approach to assess the residual error. When the residual error no longer decreases, the training is stopped and the weights are determined. Other methods of controlling both bias and variance in neural network models include reducing the number of weights by truncation or allowing a gradual increase in the network during training to prevent over-fitting.

Recurrent networks solve many of the problems associated with over-fitting and training large amounts of data in modeling dynamic processes. The inclusion of preliminary information about the process that is modeled by a neural network can allow an even greater reduction in the number of parameters.

1.9   ANN Model Validation

ANN model validation is a substantial part of system identification. Despite the presence of a large number of tests of statistical hypotheses and the fact that the evaluation criterion is developed for linear, sustainable systems, the problem is much more complex in nonlinear dynamical systems. The criterion for model validation is the value of the objective function $J(\theta, Z^N)$, when the model is applied to a data set (Z^M) different from (Z^N), which is used in system identification. However, such a criterion is not sensitive to the error obtained by model mismatch (bias) and the error due to data noise. In correlation analysis, more complex tests are performed to examine the prediction error $\varepsilon_t(\theta)$. If a nonlinear nonparametric model is adequate and unbiased, then the error must be uncorrelated with all linear and nonlinear combinations of past inputs and outputs. This output can be determined using a normalized cross-correlation function:

$$\widehat{\Phi}_{\Psi_1\Psi_2}(\tau) = \frac{\sum\limits_{t=1}^{N-\tau} \Psi_1(t)\Psi_2(t-\tau)}{\sum\limits_{t=1}^{N} \Psi_1^2(t) \sum\limits_{t=1}^{N} \Psi_2^2(t)} \tag{6.19}$$

where $\widehat{\Phi}_{\Psi_1\Psi_2}$ is the normalized cross-correlation between two variables (time series) Ψ_1 and Ψ_2, τ is a time shift, and t is a time index. $\widehat{\Phi}_{\Psi_1\Psi_2}(\tau)$ can be drawn as a function of τ for positive and negative time lags.

Figure 6.8 shows an example for $\widehat{\Phi}_{\Psi_1\Psi_2}(\tau)$, when Ψ_1 represents a sequence of white noise and shows a slight autocorrelation. Due to the fact that the evaluation correlations will never be exactly zero, the selected confidence interval is approximately 95%, as $\pm 1.96/\sqrt{N}$ for a large N to indicate if the correlations are significant. For multivariate nonlinear models, it is impossible to check every possible cross-correlation, but auto- and cross-correlation must be calculated for the residues as a minimum validation of the model.

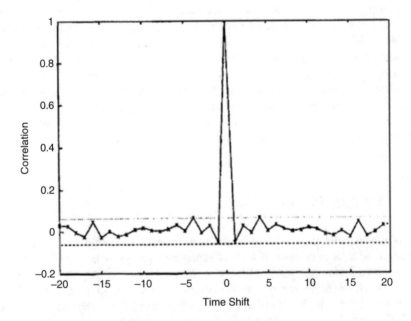

Fig. 6.8 Autocorrelation function for white noise

2 Concluding Remarks

Over the last decades, the popularity of the artificial neural network approach has been growing and it has been applied with great success in various fields of industry such as petrochemical, oil and gas industry, biotechnology, environmental protection, fuel and energy production, mineral industry, nanotechnology, pharmaceutical industry and polymers production, telecommunications, etc. The reason for this is that they show very good results in modeling complex, nonlinear production processes and systems in which traditional models have difficulties or even impossible to develop. The aim of this chapter was to present the main characteristics of artificial neural networks, the choice of their architecture, the process of training and validation of models, as well as several types of neural networks that are applied most effectively to solve real chemical problems with their advantages and disadvantages.

References

1. A.E. Rodrigues, M. Minceva, Modelling and simulation in chemical engineering: Tools for process innovation. Comput. Chem. Eng. **29**(6), 1167–1183 (2005)
2. J.D. Hoffman, S. Frankel, *Numerical Methods for Engineers and Scientists* (Marcel Dekker, New York, 2001)
3. M. Shacham, N. Brauner, Preventing oscillatory behavior in error control for ODEs. Comput. Chem. Eng. **32**(3), 409–419 (2008)
4. R.E. Precup, S. Preitl, E.M. Petriu, J.K. Tar, M.L. Tomescu, C. Pozna, Generic two-degree-of-freedom linear and fuzzy controllers for integral processes. J. Frank. Instit. **346**(10), 980–1003 (2009)
5. W.J. Cole, K.M. Powell, T.F. Edgar, Optimization and advanced control of thermal energy storage systems. Rev. Chem. Eng. **28**(2-3), 81–100 (2012)
6. Y. Meng, B.L. Lin, A feed-forward artificial neural network for prediction of the aquatic ecotoxicity of alcohol ethoxylate. Ecotoxicol. Environ. Saf. **71**(1), 172–186 (2008)
7. S. Curteanu, Different types of applications performed with different types of neural networks. In: Kwon SJ, editor. Artificial Neural Networks. Chapter 3. New York: Nova Science Publishers, 2011
8. F.S. Hoseinian, B. Rezai, E. Kowsari, M. Safari, A hybrid neural network/genetic algorithm to predict Zn(II) removal by ion flotation. J. Separat. Sci. Technol. **55**(6), 1197–1206 (2020)
9. S.M. Mousavi, E.S. Mostafavi, P. Jiao, Next generation prediction model for daily solar radiation on horizontal surface using a hybrid neural network and simulated annealing method. Energy Convers. Manag. **153**, 671–682 (2017)
10. M.A. Ahmadi, S. Zendehboudi, A. Lohi, A. Elkamel, I. Chatzis, Reservoir permeability prediction by neural networks combined with hybrid genetic algorithm and particle swarm optimization. Geophys. Prospect. **61**(3), 582–598 (2013)
11. S. Bahrami, F.D. Ardejani, E. Baafi, Application of artificial neural network coupled with genetic algorithm and simulated annealing to solve groundwater inflow problem to an advancing open pit mine. J. Hydrol. **536**, 471–484 (2016)
12. H. Cartwright, *Using Artificial Intelligence in Chemistry and Biology* (CRC Press, Taylor & Francis Group, UK, 2008)
13. F.A.N. Fernandes, L.M.F. Lona, Neural network applications in polymerization processes. Braz. J. Chem. Eng. **22**(3), 323–330 (2005)

14. B.M. Wilamowski, Neural network architectures and learning algorithms. IEEE Ind. Electron. Mag. **3**(4), 56–63 (2010)
15. S. Curteanu, Direct and inverse neural network modeling in free radical polymerization. Open Chemistry **2**(1), 113–140 (2004)
16. G. Lisa, W.D. Apreutesei, S. Curteanu, C. Lisa, C.G. Piuleac, V. Bulacovschi, Ferrocene derivatives thermostability prediction using neural networks and genetic algorithms. Thermochim. Acta **521**(1-2), 26–36 (2011)
17. J. Lobato, P. Cañizares, M. Rodrigo, C. Piuleac, S. Curteanu, J. Linares, Direct and inverse neural networks modelling applied to study the influence of the gas diffusion layer properties on PBI-based PEM fuel cells. Int. J. Hydrog. Energy **35**, 7889–7897 (2010)
18. N. Vaklieva-Bancheva, R. Vladova, E. Kirilova, Simulation of heat-integrated autothermal thermophilic aerobic digestion system operating under uncertainties through artificial neural networks. Chem. Eng. Trans. **76**, 325–330 (2019a)
19. E.G. Shopova, N.G. Vaklieva-Bancheva, Basic – A genetic algorithm for engineering problem solution. Comput. Chem. Eng. **30**(8), 1293–1309 (2006)
20. N. Vaklieva-Bancheva, R. Vladova, E. Kirilova, Methodology for energy efficiency and sustainability improvement of batch production systems on the example of autothermal thermophilic aerobic digestion systems. J. Ecol. Eng. **20**(9), 103–115 (2019b)
21. J.L. Elman, Finding structure in time. Cogn. Sci. **14**, 179–211 (1990)
22. F. Caliman, S. Curteanu, C. Betianu, M. Gavrilescu, I. Poulios, Neural networks and genetic algorithms optimization of the photocatalytic degradation of alcian blue 8GX. J. Adv. Oxidat. Technol. **11**(2), 316–326 (2008)
23. E.N. Drăgoi, S. Curteanu, F. Leon, A.I. Galaction, D. Cascaval, Modeling of oxygen mass transfer in the presence of oxygen-vectors using neural networks developed by differential evolution algorithm. Eng. Appl. Artif. Intell. **24**(7), 1214–1226 (2011)
24. E. Kirilova, S. Yankova, B. Ilieva, N. Vaklieva-Bancheva, A new approach for modeling the biotransformation of crude glycerol by using NARX ANN. J. Chem. Technol. Metall. **49**(5), 473–478 (2014)
25. Y. Chen, D.M. Himmelblau. Determination of nonlinear dynamic model order by false nearest neighbor method. In World Congress on Neural Networks, Washington, DC, July 17-21, 1995
26. J. Lee, C. Yum, W. Kim. Neural network based judgmental adjustment for time series forecasting. in EANN Helsinbu, August 1995, pp. 229
27. Z. Ahmad, J. Zhang, Selective combination of multiple neural networks for improving model prediction in nonlinear systems modeling through forward selection and backward elimination. Neurocomputing **72**, 1198–1204 (2009)
28. Z-H Zhou, J. Wu, W. Tang. Ensembling neural networks: Many could be better than all. Artif. Intell., vol. 137(1-2), pp. 239-263, 2002
29. M.H. Nguyen, H.A. Abbass, R.I. McKay, Stopping criteria for ensemble of evolutionary artificial neural networks. Appl. Soft Comput. **6**(1), 100–107 (2005)
30. A. Mukherjee, J. Zhang, A reliable multi-objective control strategy for batch processes based on bootstrap aggregated neural network models. J. Process Control **18**(7-8), 720–734 (2008)
31. F. Herrera, J. Zhang, Optimal control of batch processes using particle swam optimisation with stacked neural network models. Comput. Chem. Eng. **33**, 1593–1601 (2009)
32. J. Torres-Sospedra, C. Hernandez-Espinosa, M. Fernandez-Redondo, Combining MF networks: A comparison among statistical methods and stacked generalization artificial neural networks in pattern recognition, in *Springer*, ed. by F. Schwenker, S. Marinai, (Lecture Notes Comput Sci, Berlin/Heidelberg, 2006), pp. 210–220
33. Y. Tian, J. Zhang, A.J. Morris, Modeling and optimal control of a batch polymerization reactor using a hybrid stacked recurrent neural network model. Ind. Eng. Chem. Res. **40**, 4525–4535 (2001)
34. C. Piuleac, M. Rodrigo, P. Cañizares, S. Curteanu, C. Sáez, Ten steps modelling of electrolysis processes by using neural networks. Environ. Model. Softw. **25**(1), 74–81 (2010)
35. F. Leon, C.G. Piuleac, S. Curteanu, Stacked neural network modeling applied to the synthesis of polyacrylamide-based multicomponent hydrogels. Macromol. React. Eng. **4**(9-10), 591–698 (2010)

Chapter 7
Approach for Parameter Identification of Multiparameter Models

P. Popova-Krumova and Christo Boyadjiev

Abstract A very important stage of model development is parameter identification through inverse problem solutions.

The kinetics of many chemical, biochemical, photochemical, and catalytic reactions is very complex and as a result the kinetic model consists of many equations and parameters. Model parameter identification in these cases is very difficult because of the multiextremal least square function or because of the fact that some minima are of ravine type. The iterative solution of this problem needs very good initial value approximations for the parameters (in the attraction area of the global minimum) for the minimum searching procedure.

A polynomial approximation of the experimental data permits to propose a new hierarchical approach for obtaining initial parameters values in the global minimum area, using a consecutive approximations method.

This approach for parameter identification of multiobjective models is tested for two bioprocesses-modeling of fermentation systems and red microalgae growth kinetics. The model parameter values are obtained on the bases of real experimental data. The results obtained show a decrease of the model error variance on every next hierarchical level and a good agreement with the experimental data on the last level.

Key words Multiobjective model · Model parameter identification · New hierarchical approach · Fermentation systems

P. Popova-Krumova (✉) · C. Boyadjiev
Institute of Chemical Engineering, Bulgarian Academy of Sciences, Sofia, Bulgaria
e-mail: p.krumova@iche.bas.bg

© The Author(s), under exclusive license to Springer Nature Switzerland AG 2022
C. Boyadjiev (ed.), *Modeling and Simulation in Chemical Engineering*, Heat and
Mass Transfer, https://doi.org/10.1007/978-3-030-87660-9_7

1   Introduction

Bioprocess technology is an extension of ancient techniques for developing useful products by taking advantage of natural biological activities. Bioprocesses have become widely used in several fields of commercial biotechnology, such as production of enzymes, antibiotics and biofuels production.

The specificity of bioprocess is their modeling and model parameters identification, since they contain a large number of model equations, and hence many parameters.

The parameters identification problem for complex kinetic models was described [1, 2, 3] on the base of least square function minimization, using experimental data or its spline approximations for model equations solutions.

Model parameters identification in these cases is very difficult to be done, because of the multiextremal least square function or because of the fact that some minima are ravine type. The solution of this problem needs very good initial values of the parameters (in the global minimum area) for the minimum searching procedure. This is the main problem in the multiextremal function minimization which will be solved on the base a new hierarchical approach.

1.1   New Hierarchical Approach for Parameter Identification

Let's consider the following multiparameter model:

$$\frac{dc_i}{dt} = F_i(c_1, \ldots, c_M; k_{i1}, \ldots, k_{i\alpha_i}), \quad c_i(0) = c_{(0)\,i}, \quad i = 1, \ldots, M, \qquad (7.1)$$

where t is the time, $c_i(t_n)$ and k_{ij} are objective functions (concentrations of the reagents) and parameters in the model for $i = 1, \ldots, M, j = 1, \ldots, \alpha_i$, α_i—number of parameters of the ith equation.

For parameter identification problem solution will be used experimental data for the objective functions: $c_i^{(e)} = c_i^{(e)}(t_n)$, $n = 1, \ldots, N$, where N is the number of experimental data measurements.

The least square functions for the separate model equations are

$$Q_i(k_{i1}, \ldots, k_{i\alpha_i}) = \sum_{n=1}^{N} \left[c_i(t_n) - c_i^{(e)}(t_n) \right]^2, \quad i = 1, \ldots, M, \qquad (7.2)$$

where $c_i(t_n)$, $i = 1, \ldots, M$ are obtained after solution of (7.1). The least square function of the parameter identification in the model (7.1) is: $Q = \sum_{i=1}^{M} Q_i..$

The total parameters number $I = \sum\limits_{i=1}^{M} \alpha_i$ is very large and in many cases it is not possible to minimize function Q, because this function is multiextremal and some of minima are ravine types.

The minimization of Q is multicriterial optimization problem with equal specific weight coefficients of the separate criteria. The obtaining of the global minimum point needs of very well initial approximation, i.e., the initial point of the minimization procedure has to be in the global minimum area.

The experimental data for the objective functions (concentrations) can be presented, using polynomial approximations:

$$c_i^{(e)}(t_n) \rightarrow P_i(t), \quad n = 1, \ldots, N, \quad i = 1, \ldots, M, \tag{7.3}$$

where $P_i(t)$ are polynomials of 5 or 6 power.

Let's consider the first equation of (7.1), where all functions, including the first, are replaced by the polynomial approximations of the experimental data: $c_i(t) \equiv P_i(t), \quad 2 \leq i \leq M,$.i.e.,

$$\frac{dc_1}{dt} = F_1(c_1, P_2, \ldots, P_M; k_{11}, \ldots, k_{1\alpha_1}), \quad c_1(0) = c_{(0)\,1}. \tag{7.4}$$

The minimization of the least square function

$$Q_1(k_{11}, \ldots, k_{1\alpha_1}) = \sum_{n=1}^{N} \left[c_1(t_n) - c_1^{(e)}(t_n) \right]^2, \tag{7.5}$$

where $c_1(t_n)$ is the solution of the eq. (7.4), permit to calculated parameters values $\left(\tilde{k}_{11}, \ldots, \tilde{k}_{1\alpha_1} \right)$ in the first equation in the model (7.1). The parameters identification errors for the eq. (7.4) are result of the experimental data errors and polynomial approximations errors only, i.e., it has not an influence of all another model equations ($2 \leq i \leq M$) errors.

This procedure is possible to repeat for all eqs. (7.1) (step by step). The obtained parameter values $\tilde{k}_{ij}; i = 1, \ldots, M; j = 1, \ldots, \alpha_i$ are possible to be used as a zero order approximation for the model parameter identification and they are the zero hierarchical level in the parameter identification procedure.

Very often the exactness of the zero order parameters approximations is not sufficient for the minimization of Q, because the parameters in one model do not give an account of another models error. The first order approximation may be obtained using the zero order one. It will be made on the first hierarchical level (step by step).

The first step is consecutive solutions of the model equations and to obtain the first order approximation of the parameters.

The last step is solution of i equations

$$\frac{dc_1}{dt} = F_1\left(c_1, \ldots, c_i; \tilde{k}_{11}, \ldots, \tilde{k}_{1a_1}\right), \quad c_1(0) = c_{(0)\,1},$$

$$\frac{dc_i}{dt} = F_i(c_1, \ldots, c_i; k_{i1}, \ldots, k_{ia_i}), \quad c_i(0) = c_{(0)\,i}, \quad i = 2, \ldots, M, \qquad (7.6)$$

and to obtain $\left(\overline{k}_{i1}, \ldots, \overline{k}_{ia_i}\right)$ after minimization of Q_i.

The last hierarchical level is to solve the eqs. (7.1) and to minimize Q using the first order approximation values $\overline{k}_{ij}(i = 1, \ldots, M; j = 1, \ldots, a_i)$ as initial approximations of the minimization procedure.

The hierarchical approach for parameter identification of multiobjective models is tested for a fermentation system and microalgae growth kinetics modeling, using the real experimental data [4, 5].

1.2 Fermentation Systems Modeling

The mathematical models of the fermentation systems contain biomass, product and substrates material balances. The obtained models [4, 5] consist of 3–4 equations with 6–10 parameters, which have to be obtained using experimental data [4].

The use of the new hierarchical approach to solve multiobjective models permits to obtain the exact parameter values and the results obtained are in a good agreement with the experimental data (Figs. 7.1, 7.2, 7.3, and 7.4).

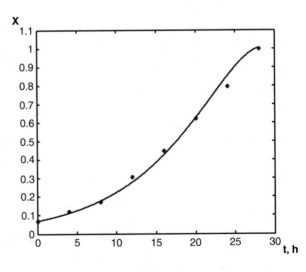

Fig. 7.1 Comparison of the calculated values and experimental data for biomass dimensionless concentration

Fig. 7.2 Comparison of the calculated values and experimental data for gluconic acid dimensionless concentration

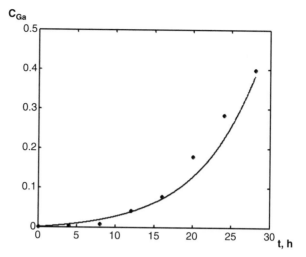

Fig. 7.3 Comparison of the calculated values and experimental data for glucose dimensionless concentration

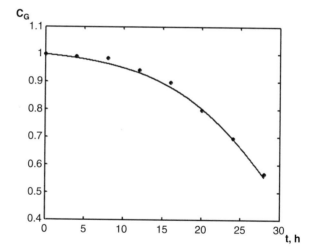

1.3 Microalgae Growth Kinetics Modeling

The use of the hierarchical approach [6] to solve multiobjective models is impossible in the case of *incomplete experimental data*, i.e., the experimental data for concentration–time dependences of some reagents or reaction products are missing. An example of this case is modeling of microalgae growth kinetics [7].

The mathematical model of the process is characterized by four parameters that will be obtained from the experimental data.

Fig. 7.4 Comparison of the calculated values and experimental data for oxygen dimensionless concentration

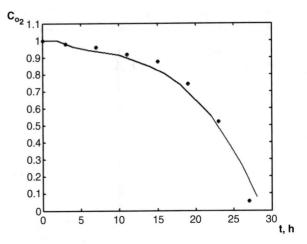

$$\frac{dc_X}{dt} = \mu_{max} \frac{c}{k_1 + c} c_X - k_0 c_X$$

$$\frac{dc}{dt} = Q - A_x \mu_{max} \frac{c}{k_1 + c} c_X$$

$$Q = k \left(\frac{c_{0,gas} + c_{h,gas}}{2} - k_H c \right), \quad c_{h,gas} = \frac{\left(\frac{u}{h} - \frac{k}{2} \right) c_{0,gas} + k k_H c}{\frac{u}{h} + \frac{k}{2}} \tag{7.7}$$

$$t = 0, \quad c_X = c_{X_0}, \quad c = c_0$$

The experimental data for the biomass concentration will be representing by the polynomial approximation:

$$c_X(t) = P(t), \quad \frac{dc_X}{dt} = \frac{dP}{dt} = P'(t), \quad A(t) = \frac{P'(t)}{P(t)} \tag{7.8}$$

The lack of experimental data for the concentration $c(t)$ will be substituted by the "experimental data" $\widehat{c}(t)$:

$$\widehat{c}(t) = \frac{k_1 [k_0 + A(t)]}{\mu_{max} - k_0 - A(t)}. \tag{7.9}$$

For identification of the model parameters, the least square function has to be used:

$$F = \sum_{i=1}^{N} [c_X(t_i) - c_{X_{exp}}(t_i)]^2 + a \sum_{i=1}^{N} [c(t_i) - \widehat{c}(t_i)]^2. \tag{7.10}$$

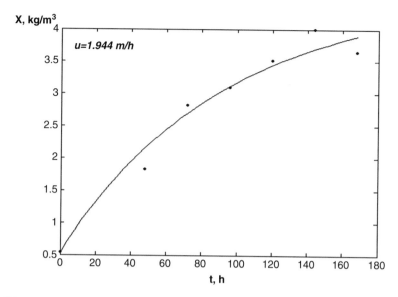

Fig. 7.5 Comparison of the calculated values and experimental data for biomass concentration

On the basis of the existence of a relationship between the concentration of biomass and the CO_2 concentration in the liquid phase, it is demonstrated that the comparison of the theoretical and experimental data in Figs. 7.5 and 7.6 show that the accuracy of the solution could be increased provided more detailed experimental data for the substitution of missing experimental data with a "provisional experimental data set" are available. The last set depends on the model parameters.

Comparisons between the model, with calculated parameters, and the experimental data are shown on Figs. 7.5 and 7.6.

2   Conclusions

The proposed method solves parameter identification problem for multiparameter models, when the least square function is multiextremal.

The using of polynomial approximations for the experimental data permits to obtain parameter values in separate model equations. The method is tested on fermentation process and microalgae growth kinetic modeling. The results obtained are in a good agreement with the experimental data.

The proposed parameter identification approach offers a possibility for finding a solution of some problems connected with multiparameter models and in the cases of insufficient experimental information. The method presented is applicable for different fermentation and photosynthetic processes.

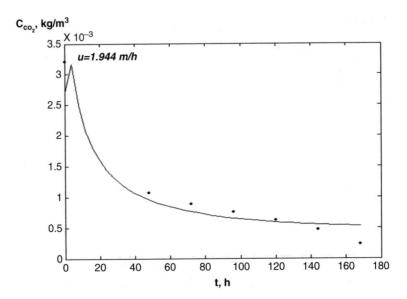

Fig. 7.6 Comparison of the calculated values and experimental data for CO_2 concentration in the liquid phase

Acknowledgement This work is supported by Project of Fundamental Scientific Research 19-58-18004, conducted by RFBR and the National Science Fund of Bulgaria under contract No KP 06 RUSIA-3 from 27 Sep. 2019, "Modeling, simulation and experimental investigations of the interphase mass transfer and separation in distillation, absorption, adsorption and catalytic processes in industrial column apparatuses" and contract No KP 06-H37/11/ 06.12.2019 "Integrated absorption-adsorption process for waste free decontamination of gases from sulfur dioxide."

References

1. J. Madar, J. Abonyi, H. Roubos, F. Szeifert, Incorporating prior knowledge in a cubic spline ApproximationsApplication to the identification of reaction kinetic models. Ind. Eng. Chem. Res. **42**, 4043–4049 (2003)
2. D.M. Himmelblau, C.R. Jones, K.B. Sichoff, Determination of rate constants for kinetics models. Ind. Eng. Chem. Fundam. **6**(4), 539–543 (1967)
3. Y.P. Tang, On the estimation of rate constants for complex kinetics models. Ind. Eng. Chem. Fundam. **10**(2), 321–322 (1971)
4. V. Beschkov, S. Velizarov, Oxygen transfer and glucose to gluconic acid oxidation by cells of Gluconobacter oxydans NBIMCC 104. Comptes Rendus Acad. Bulg. Sci. **47**(8), 53–57 (1994)
5. J. Merchuk, M. Gluz, I. Mukmenev, Comparison of photobioreactors for cultivation of the red microalgae Porphyridium sp. J. Chem. Technol. Biotechnol. **75**, 1119–1126 (2000)
6. P. Popova, C. Boyadjiev, Hierarchical approach for parameter identification of multiparameter models. Biochem. Eng. J. **39**, 397–402 (2008)
7. P. Popova, C. Boyadjiev, On the red microalgae growth kinetics modeling. Chem. Biochem. Eng. Q. **12**, 491–498 (2008)

Chapter 8
Modeling and Simulation of Phase Change Material Based Thermal Energy Accumulators in Small-Scale Solar Thermal Dryers

J. Patel, J. Andharia, A. Georgiev, D. Dzhonova, S. Maiti, T. Petrova, K. Stefanova, I. Trayanov, and S. Panyovska

Abstract Solar thermal energy is of intermittent and dynamic character and the necessity to use this energy during non-sunshine periods has led to the development of thermal energy accumulators. The need of compact solutions have prompted researchers towards using latent heat storage. Phase change materials as thermal energy storage are attractive because of their high storage density and characteristics to release thermal energy at constant temperature corresponding to the phase transition temperature. The chapter overviews the recent state-of-the-art small-scale solar thermal dryers integrated with phase change material as an energy accumulator. This is an intensive field of investigation for more than 30 years with importance for the agriculture and the food industry especially in hot climate. A variety of commercial small-scale solar dryers are offered as a low-cost, zero-energy solution for small farmers. And yet, there are no commercial systems using latent thermal storage because at the present level of development this unit will increase unacceptably the price of the system. The solution needs very simple design, accessible materials, and optimal conditions for operation.

The aim of the present work is to make an overview of the methods for theoretical evaluation and prediction, which are used to design and assess this devices and to point out the most appropriate of them for this new solution. The models enable to distinguish the most cost- and energy-effective solar dryer systems with thermal storage among the great number of designs, devices, and materials. The resulting conclusions from the collected and compared information will serve as a base for a

J. Patel · J. Andharia · S. Maiti
CSIR-Central Salt & Marine Chemicals Research Institute, Bhavnagar, Gujarat, India

A. Georgiev
Institute of Chemical Engineering at the Bulgarian Academy of Sciences, Sofia, Bulgaria

Department of Mechanics, Technical University of Sofia, Plovdiv Branch, Plovdiv, Bulgaria

D. Dzhonova (✉) · T. Petrova · K. Stefanova · I. Trayanov · S. Panyovska
Institute of Chemical Engineering at the Bulgarian Academy of Sciences, Sofia, Bulgaria
e-mail: dzhonova@bas.bg

© The Author(s), under exclusive license to Springer Nature Switzerland AG 2022
C. Boyadjiev (ed.), *Modeling and Simulation in Chemical Engineering*, Heat and Mass Transfer, https://doi.org/10.1007/978-3-030-87660-9_8

novel solution of a cost-effective thermal energy storage for a small-scale solar dryer, which will lead to improved efficiency of the drying process, due to controlled temperature and longer operational time. This information might serve also in the development of the wider field of thermal energy storage, which is an important part of the technologies of renewable and waste energy conversion.

Key words Computational Fluid Dynamics · Energy efficiency · Exergy efficiency · Latent heat · Phase change material · Solar dryer · Thermal energy accumulator

Nomenclature

A	Area, m^2
c_p	Specific heat, J/(kg K)
$c_{p,SAH}$	Average specific heat of air between $T_{SAH,i}$ and $T_{SAH,o}$, J/(kg K)
$c_{PCM,s}$	Average specific heat of solid PCM J/(kg K)
$c_{PCM,l}$	Average specific heat of liquid PCM, J/(kg K)
E	Energy, J
Ex	Exergy, J
$\dot{Ex}$	Exergy flow rate, W
g	Gravitational acceleration vector, m/s^2
h	Sensible enthalpy, J/kg
h_{fg}	Latent heat of vaporization, J/kg
H	Enthalpy, J/kg
I	Solar intensity, W/m^2
k	Thermal conductivity, W/(mK)
L	Heat of fusion per unit mass, J/kg
m	Mass, kg
$\dot{m}$	Mass flow rate, kg/s
p	Static pressure, Pa
$\overline{P}$	Time-averaged pressure, Pa
P_{fan}	Power consumption of fan, W
Pr	Prandtl number
R	Gas constant, J/(kgK)
Q	Thermal energy, J
$\dot{Q}$	Heat flow rate, W
t	Time, s
T	Temperature, K
$\overline{T}$	Time-averaged temperature, K
$\overline{T'}$	Temperature fluctuation, K
$\overline{u_i}$	Time-averaged velocity component, m/s
$\overline{u_i'}$	Velocity fluctuation, m/s
x_i	Coordinate axis, m
v	Velocity vector, m/s

Greek Letters

α	Convective heat transfer coefficient, W/(m^{2}K)
α'	Absorptivity
η	Thermal efficiency
β	Liquid volume fraction
γ	Thermal expansion coefficient, 1/K
Δ	Difference
μ	Dynamic viscosity, Pa.s
ρ	Density, kg/m^3
τ	Stress tensor, Pa
τ'	Transmissivity

Subscripts

a	Air
abs	Absorber
ch	Charging
d	Dryer
dis	Discharging
des	Destruction
e	Evaporated moisture
es	Energy storage
f	Fluid
f_{ch}	Final in charging
f_{dis}	Final in discharging
F	Fusion
i	Inlet
i_{ch}	Initial in charging
i_{dis}	Initial in discharging
in	Input
l	Liquid
o	Outlet
out	Output
PCM	Phase change material
r	Reference
re	Received
s	Solid
sys	Drying system
SA	Solar accumulator
SAH	Solar air heater
w	Wax

Abbreviations

BC	Boundary conditions
CFD	Computational fluid dynamics
DC	Drying chamber
ETC	Evacuated tube collector
FLT	First law of thermodynamics
FPC	Flat plate collector
HDPC	High density polyethylene containers
HE	Heat exchanger
HTF	Heat transfer fluid
LHS	Latent heat storage
PCM	Phase change material
SAH	Solar air heater
SLT	Second law of thermodynamics
TES	Thermal energy storage

1 Introduction

Solar energy is one of the most prospective sources of renewable energy. Its intermittent nature is a drawback which can be overcome to some extent by energy storage for energy supply during the non-sunshine hours. Thermal energy storage (TES) is one of the useful solutions. It includes three principles of heat accumulation, by using sensible heat, latent heat, and chemical heat. Intensive research is carried out recently on thermal energy storage in combination with a solar dryer (Fig. 8.1). The wide interest in these systems is due to the increasing need for preservation of food at low energy consumption with high quality of the product. The heat storage diminishes the fluctuation in the inlet temperature and supplies heat flow near constant temperature. The constant temperature regulation helps to prevent degradation of product quality. The environmental effect is reduced harmful emissions in the atmosphere.

1.1 Sensible Heat Versus Latent Heat in TES for Integration with a Solar Dryer

In sensible heat storage, the material rises its temperature. The amount of heat stored depends on the density, volume, specific heat, and temperature difference of the storage material. The latent heat is stored when the storage material changes its phase at the working temperature. The advantage of the latent heat storage (LHS) in respect to sensible one is higher energy storage density, i.e., much smaller mass and volume

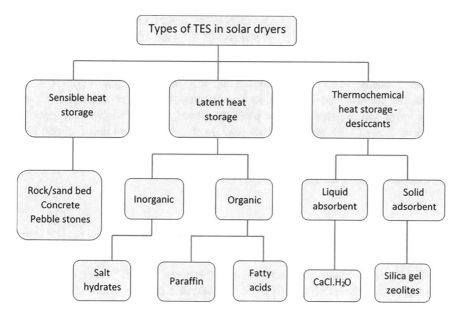

Fig. 8.1 Types of TES integrated in solar dryer systems

of the storage material is needed to store a certain amount of energy. This is illustrated by the example given in [1] that "in the case of water, 80 times as much energy is required to melt 1 kg of ice as to raise the temperature of 1 kg of water by 1 °C."

The rock bed is the most common material for sensible storage used in solar dryer systems. A simplified solar energy storage of a rock bed was investigated [2] for drying of peanuts. The drying time ranged from 22 to 25 h reducing the moisture content to the safe moisture level of 20% with an air flow rate of 4.9 m^3/s.

It should be noted that the high energy storage density of the phase change material (PCM) is an advantage, but does not guarantee higher efficiency of the system. Surprising is the conclusion in [3] that using sensible heat storage with pebble stones may be more advantageous for the drying process than LHS with paraffin wax, in regard to performance and cost. The study presented experiments for comparison of two systems for heat storage in a solar air dryer. One of the systems was a packed bed of pebble stones and the other was paraffin wax with a melting temperature of 55–60 °C, placed in heat resistant bags. One and the same quantity of 500 kg of each material was used to dry 10 kg of 5 mm thick lemon slices. The drying process lasted an average of 6.23 h (paraffin) and 6.27 h (pebble stones). The total discharge time of the two systems was 7 h on average. The presented data demonstrate the importance of the design of the LHS. The advantage of the PCM can be wasted because of hindered air flow and poor heat transfer provided by the bags containing the PCM.

The chemical heat storage is characterized by highest energy storage density of the material, but this technology is less mature than the previous two in respect to commercialization [4]. It uses reversible reactions which involve absorption and release of heat. In solar dryers, an example of this technology is liquid or solid desiccant material used for reducing the humidity of the drying air in off-sunshine hours. The performance of an indirect forced-convection, desiccant-integrated solar dryer for drying of green peas and pineapple slices was studied [5]. The desiccant was a mixture of 60% bentonite, 10% calcium chloride, 20% vermiculite, and 10% cement. It was molded in the shape of cylinders. In this process, the moisture was removed from the drying air by adsorption in the desiccant material which was regenerated by solar energy.

1.2   Conditions for Successful Solar Drying

Drying under controlled conditions of temperature and humidity helps the different products to dry reasonably rapidly to a safe moisture content and to ensure a superior quality [6–8]. The temperature range of 40–60 °C is found to be sufficient for most of the food products to retain their flavor, aroma, texture, and nutrition values. Therefore, the minimum melting temperature of the PCM should be 5–10 °C higher than the desired temperature of the heat transfer fluid (HTF) [9]. The good design of a solar dryer has to ensure continuously a stable temperature of the drying air in the order of 10–25°C above the ambient temperature [10]. This is necessary to avoid the dried object reabsorbing the moisture during night, when at absence of solar energy, the air temperature drops and its humidity increases. This will make the drying time longer and in the worst case, it can ruin the dried object because of mold [11, 12]. Initial and final moisture content and maximum allowable temperature for drying of some crops are summarized in [13]. An overview of solar drying of medicinal plants and herbs is presented in [14] to show the drying conditions for good quality of heat sensitive products.

The focus of the present work is on integration of a solar dryer with latent thermal energy storage to prevent the interruption of the drying process during night. Using PCMs with solar dryers leads to reduction of the heat loss and the mismatch between supply and demand of thermal energy, as well as to improvement of the energy efficiency of the system [15].

The aim of the present overview is to reveal the current level of development of LHS in solar air dryers, which is an area of intensive research offering many engineering solutions, experimental data, and mathematical approaches for prediction and evaluation of the systems. The particular task is to find the appropriate methods for optimization of the thermal behavior of the LHS with a focus on paraffin as PCM for cost- and energy-efficient design of solar dryers of small capacity (for households and small producers). The most important conditions which will ensure cost- and energy-efficient system are identified on the base of the available solutions. The characteristics of a thermal energy storage, proper for integration with a small-scale solar dryer are revealed. Modeling tools are described, which can be used to improve the design of a low-cost and zero-energy drying system. The increasing of the costs of the dryer, when integrating a TES unit, with the currently available constructions and materials, is still unsolved issue, which hinders commercialization. The present work is part of the efforts for more effective utilization of renewable energy sources, which will lead to higher yields of product per device at lower ecological risk combined with a lower cost of the devices.

2 Small-Scale Solar Dryers with Paraffin

It has been pointed out in [16] that the development of a latent heat thermal energy storage system involves the understanding of three essential subjects: PCM, containers' material and design of the heat exchangers (HEs). Solid–liquid PCMs have proved to be economically attractive for use in thermal energy storage systems. Paraffin waxes have been distinguished [17] among LHS materials due to their availability in a wide temperature range, but the major problem is their low thermal conductivity. The conditions in the HE of the solar dryer with paraffin (material and surface area, insulation, air flow rate and velocity, type of the air channels, grids for food) may compensate for this shortcoming. The solutions appropriate for a simplified small-size solar dryer with paraffin are discussed in the present chapter.

2.1 Why Paraffin as PCM in Solar Dryers

Paraffin wax has found wide application as LHS in solar dryers due to its advantages: high density of energy storage, ability to operate in a variable range of temperature conditions, long-term storage without heat loss (i.e., no self-discharge), constant temperature of phase change, no super cooling, low vapor pressure, good thermal and chemical stability and self-nucleating behavior, low cost. The heat is stored, when paraffin changes its phase from solid to liquid, and released at a constant temperature, when paraffin cools down and solidifies again. Paraffins are tabulated in [16] with their number of carbon atoms (from 14 to 34) and categorized as more or less promising according to their characteristics. Paraffin consists of a mixture of n-alkanes CH_3-(CH_2)-CH_3 into which the crystallization of the (CH_3) - chain is responsible for a large amount of energy absorption. The melting point and heat of fusion increase with molecular weight. The latent heat of fusion of paraffin varies from nearly 170 kJ/kg to 270 kJ/kg with temperature of phase change between 5 and 76°C [16], which makes them suitable for building and solar applications. Paraffin is safe, reliable, predictable, less expensive, and non-corrosive. Major drawbacks are low thermal conductivity, possible non-compatibility with the plastic container, and moderate flammability [16]. The following recent studies on paraffin wax in solar air dryers show the important preferred characteristics for that application.

Five paraffin waxes and wood resin were compared by studying their thermophysical properties [18]. The investigation aimed at selection of PCM for use as heat storage in a solar dryer. The wood resin was rejected. The selected paraffin had density of 932.9 kg/m^3 in liquid state, and maximum latent heat of fusion and solidification 383.87 kJ/(kg K) and 320.26 kJ/(kg K), respectively. The selected PCM was used in the flat plate collector (FPC) of the solar dryer to identify the thermal zones and to validate its capability as thermal storage. The maximum temperature achieved at the FPC outlet was 50 °C. It was found that after 18:00, the average temperature of the collector chamber with the selected paraffin was 23.5% higher than that with no PCM.

Paraffin with melting point 45–48 °C was found more suitable compared to paraffin with melting point 68–70 °C in a fish dryer [19] with a maximum temperature at the inlet of 75 °C. Using PCM to control the temperature inside the chamber was an important aspect as fish could not sustain temperature more than 62°C.

2.2 Thermal Conductivity Enhancement of Paraffin in Solar Dryers

Multiple techniques (Fig. 8.2) are used to increase the charging/discharging rate of paraffin by increasing its thermal conductivity. They include dispersion of high conductivity particles in the PCM [15, 20] using extended surfaces, and embedding porous material in the PCM.

The use of finned tubes, as well as metal fibers and metal matrix, for example, resulted in an increase in the effective thermal conductivity of the PCM up to five times, leading to higher rate of the heat transfer [21].

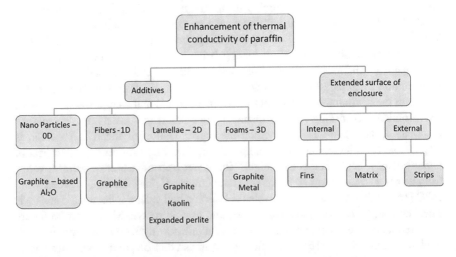

Fig. 8.2 Techniques for enhancement of thermal conductivity of paraffin

2.2.1 Additives

Common additive materials, used with paraffin and other PCM [22, 23, 24], include: carbon materials (matrix, particles, fibers, nanotubes, foam, expanded graphite, graphite powder, graphene, graphene aerogel); metal like Al, Cu, Ni (nanoparticles, foam). Some novel materials for preparation of high performance PCM are categorized as 3D, 2D, 1D, and 0D additives [25]. Examples of these groups of additives used with paraffin are: graphene-nickel foam, graphite and metal foams (3D); expanded perlite, kaolin, expanded graphite (2D lamellar structure); graphite fiber (1D); and graphite-based nanoparticles, nano-Al_2O_3 (0D). Two to ten times improvement in thermal conductivity of paraffin wax with additives is reported in [25].

2.2.2 Encapsulation

The main advantages of PCM encapsulation are providing large heat transfer area, reducing the PCM reactivity towards the outside environment and controlling the changes in the volume of the storage materials as the phase change occurs [26]. Based on size, the PCM encapsulation is classified into nano (0–1000 nm), micro (0–1000 μm), and macro (above 1 mm) encapsulation. The cheapest containers used for macro-encapsulation are tin cans and plastic bags or bottles. Typical shapes of containers for PCM are discussed in a review [27] with an emphasis on the type of the geometric configuration and orientation of the container. The shapes include spherical, rectangular, cylindrical (both horizontal and vertical), and annular containers. It has been concluded there that increasing the height/width ratio of the container of the same volume decreases the time for the melting process due to the stronger buoyancy effect. Usually the material of the shell is plastic or metal (copper, aluminum, and steel) when higher heat transfer rates are desirable. The results in [28] show that a rectangular container needs nearly half of the melting time of a cylindrical container of the same volume and heat transfer area. The duration of heat discharge increases with increasing PCM container diameter in the order of sphere, cylinder, plate, and tube [29].

Heat storage of 300 soft drink cans filled with 56 kg of paraffin wax as PCM was proposed in [30]. The cans were oriented perpendicular to the air flow, resulting in 30 rows of 10 cans each. Cans were also cut into thin strips and 4.8 kg of these aluminum strips were placed into the wax to increase the effective thermal conductivity of the PCM.

2.2.3 Extended Surfaces

One of the serious problems associated with the operation of PCM storage system is the heat transfer in and out of the element containing the PCM. Paraffin wax (phase change 35–54 °C, heat of fusion 196.05 kJ/kg) was used in a solar dryer for sweet

Fig. 8.3 A scheme of a latent heat storage vessel with fins [31]

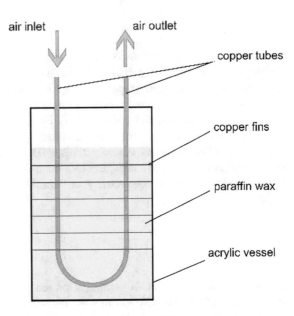

air inlet air outlet

copper tubes

copper fins

paraffin wax

acrylic vessel

potato coins [31]. The TES tank was a cylindrical acrylic vessel where the HTF (air) flowed in a tube with 18 copper fins (Fig. 8.3).

It was found that melting was dominated by heat conduction followed by free convection; charging time decreased with the increase in the inlet air temperature and the air velocity. The discharging time of the LHS at an air velocity of 1 m/s was 180 min and at an inlet velocity of 2 m/s it was 165 min. The result indicated that the air velocity did not affect much the discharging time since heat conduction was dominant during solidification. The extracted energy decreased from 1920 kJ min/kg to 1386 kJ min/kg as the inlet velocity increased from 1 to 2 m/s, respectively.

The discharging rate of a novel LHS unit [32] was significantly improved by inclusion of longitudinal fins to the tubes of the HE with paraffin as TES material.

2.3 PCM-Based Thermal Energy Accumulators in Small-Scale Solar Dryers

For solar dryers with integrated TES, the classification of [13] can be adopted as follows (Fig. 8.4).

- By working principle: direct, indirect, mixed mode, and hybrid.
- By type of construction: cabinet (box-type) and greenhouse.
- By mode of operation: passive (natural convection) and active (forced convection).

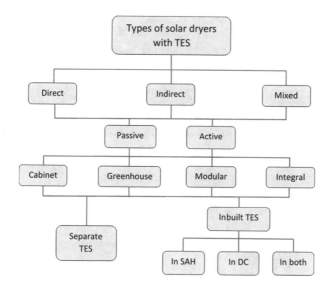

Fig. 8.4 Types of a solar dryer system with TES

– The TES can be integrated in the dryer system as: separate unit, connected outside the solar air heater (SAH); inbuilt in the SAH or in the drying chamber (DC) (at the top, bottom, side walls).

The development of a solar air drying system with paraffin includes solving several essential problems leading to energy and cost efficiency of the system: location and volume occupied by the PCM; intensification of the heat transfer between PCM and HTF; low energy consumption and heat loss of the systems.

2.3.1 Passive (Natural Convection) Versus Active (Forced Convection) Solar Drying

The heated air for drying can be driven by buoyancy forces or by a fan. The air is drawn through the DC by natural convection or by a fan, or by both. In general, the following dependencies are observed: the higher the air mass flow rate, the higher the collector efficiency; the electrical energy consumption of the fan increases with the increase in the air mass flow rate; the effect of leakages increases with the increase in the air flow rate. Natural-convection solar dryers have become more suitable for the rural sector and remote areas as they do not require external energy source. But it is generally agreed that well-designed forced-convection solar dryers are more effective and more controllable than natural circulation types. The usage of forced convection in the drying system can reduce three times the drying time and can decrease by 50% the necessary collector area [33]. Fans may be powered with

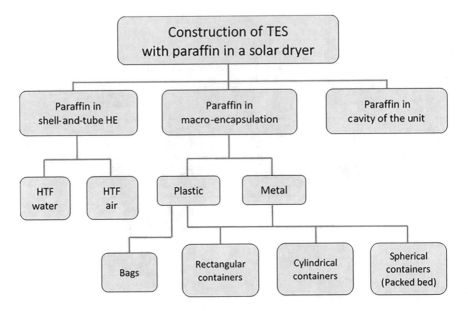

Fig. 8.5 Types of constructions of TES with paraffin in a solar dryer system

utility electricity if it is available, or with a solar photovoltaic panel. It should be noted that in the case of herbs and spices, the role of the dryer is not to dry more quickly, but to give a better-quality product.

The existing designs of TES with PCM in solar dryers are shown in Fig. 8.5. The PCM can fill part of the volume of the unit or can be encapsulated in containers with different form, size and material, arranged or dumped in a packed bed.

2.3.2 Separate TES Unit with PCM

Examples of PCM Occupying the Volume of the Unit

The performance of a PCM-based solar dryer was analyzed [34] for drying black turmeric. The drying chamber was of a mixed-mode type with hot air entering from one side and a glass top cover allowing the direct sun radiation to pass into the drying chamber. Paraffin wax was used as PCM. The TES system was a shell-and-tube HE and the paraffin wax (35 kg) was placed in the shell side (Fig. 8.6). The air from the collector passed through the tubes made of copper; one tube at the center and the other nine tubes at the periphery. The results of the solar dryer with PCM were compared to open sun drying and there was 60.7% saving of time. The discharging of paraffin wax maintained the temperature 4–5 °C higher than the ambient temperature for 6 hours after sunset.

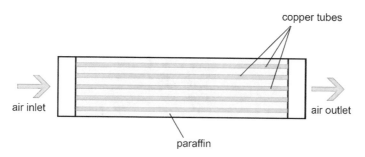

Fig. 8.6 Shell-and-tube TES [34]

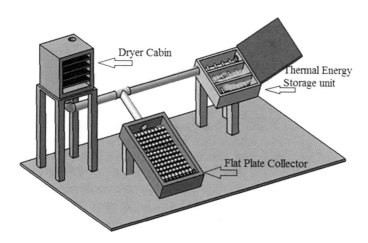

Fig. 8.7 Solar air dryer with a separate TES unit [36]

Energy and exergy analysis of solar drying of 20 kg of red chili in four consecutive sunny days (8–18 h) was performed [35]. The components of the solar dryer were two double-pass solar air heaters connected in series, a paraffin wax in shell-and-tube LHS unit, a blower, and a parallel flow DC.

Performance enhancement of a solar air dryer was focused in [36], Fig. 8.7. The solar FPC provided outlet air temperature up to 100 °C at natural convection mode during sunshine hours, when the TES was disconnected from the system. Direct solar radiation was incident on the TES unit during the charging period. At off-sunshine hours, the solar TES was connected to the DC and the FPC was disconnected. A fan was located at the base of the dryer cabin to maintain the air flow at night. The TES unit consisted of an absorber plate, a reflective mirror and heat pipes with fins as a vapor-liquid phase change device, dipped in paraffin wax as PCM, Fig. 8.7. The efficiency decreased with the increase in the air velocity. Higher temperature gradient led to an increase in the efficiency.

Fig. 8.8 Separate TES unit with PCM in HDPC [37]

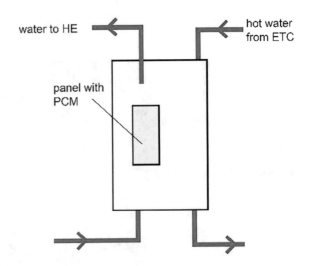

Examples of PCM Encapsulated in Containers

The suitability of two PCMs is investigated [37] for development of an active thermal storage system for a solar kiln (semi-greenhouse type) for wood drying. The storage system consisted of a water storage tank with PCM placed inside the water in high density polyethylene containers (HDPC), Fig. 8.8. The water in the tank was heated by solar energy using an evacuated tube collector (ETC) array. An additional insulation for the big north wall with 6 mm thick plywood and 100 mm thick glass-wool layer was provided. The day time solar energy was utilized to heat the water which charged the PCM. In the night, the stored heat of the PCM was utilized to keep the air inside the kiln warmer using the discharged thermal energy of the PCM. Two organic PCMs, OM 55™ (mixture of fatty acids) and paraffin wax were used. It was found that the paraffin wax had higher temperature reduction rate compared to the fatty acids.

2.3.3   TES Inbuilt in the SAH: Solutions for Better Thermo-Hydraulic Performance

The most common solution for integration of thermal storage in the SAH is to place the PCM in the volume under the absorber plate. This construction (Fig. 8.9) was adopted in the study [38]. Part of the heat from the solar radiation heated the air for the drying process and the rest of it charged the PCM. These two processes were separated in two solar air collectors in the proposed design. An indirect type forced-convection solar dryer using PCM was constructed and experimentally investigated. It consisted of two SAH and a DC. SAH 1 was a solar air panel without PCM heating

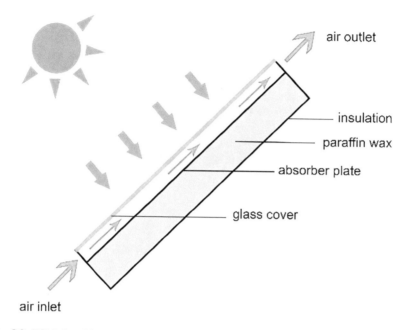

Fig. 8.9 PCM placed below the absorber plate of the SAH [38]

the air for the DC during sunshine period. SAH 2 was thermal storage with paraffin wax as PCM heating the drying air during off-sunshine period. The solar energy accumulator in [38] comprised a PCM cavity (with dimensions 2.04 m x 1.04 m and total volume of 0.33 m³) with insulator and a cover glass. Experiments were conducted in no-load conditions (without material for drying) to evaluate the charging and the discharging characteristics of the latent heat unit. The daily energy efficiency of the solar energy accumulator reached 33.9%, while the daily exergy efficiency reached 8.5%. The solar dryer of 0.7 m³ with 60 kg of paraffin wax kept the relative humidity inside the DC of 0.768 m³ between 17% and 34.5% lower than the ambient relative humidity and maintained the DC temperature 4–16 °C higher than the ambient temperature all the night.

There are different factors affecting the SAH efficiency, e.g., collector dimensions, type of absorber plate, glass cover plate, wind speed, air flow rate [39]. The absorber plate shape factor (the ratio of the total collector area to the absorber area normal to the solar radiation) is one of the most important parameters in the design of SAH. The increase in the total absorber area increases the heat transfer to the air flow, but leads to an increase in the pressure drop, therefore a higher power consumption [40]. There is an optimal range of the shape factor for maximal efficiency. The thermal storage in the SAH should be designed in such a way as not to create additional pressure drop, especially in natural convection units.

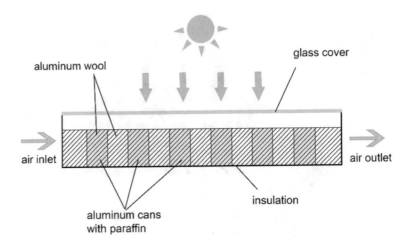

Fig. 8.10 Paraffin wax in aluminum containers inside the SAH with aluminum wool inside and outside the containers [42]

Containers with Paraffin: Reducing inside and outside Conductive Resistance

It was found [41] that the poor thermal conductivity of PCM had negligible effect on the heat transfer due to high surface convective resistance provided by the air in a packed bed storage unit for low-temperature solar air heating application. A solar air HE (Fig. 8.10) containing paraffin wax encapsulated in aluminum cans was investigated [42]. The cans were filled with 5% w/w aluminum wool, which doubled the thermal conductivity of the paraffin. The usage of aluminum wool outside the PCM containers reduced the conductive resistance of air and increased the HE efficiency in charging stage from 46.8% to 48.9% and in discharging stage from 64.4% to 80%. The authors explored the factors, which can improve the drying conditions as kiwi slicing thickness and using control systems to regulate the air flow through the devices.

Increasing the Heat Transfer Surface of the SAH with PCM

There are different solutions to increase the heat transfer surface of the air passage in the SAH. A widely employed configuration is a double-pass SAH, where the air flows over and under the absorber plate. An asymmetric double-pass air heater containing paraffin macro-encapsulated in rectangular or cylindrical metallic containers (Fig. 8.11) was studied [43].

Other solutions are corrugated absorber surface [44], fins or obstacles of different orientation and form, e.g., perforated V-blocks [45], baffles on the absorber plate [46, 47], which intensify the heat transfer by increasing the heat transfer area, creating turbulence in the air flow, and preventing from stagnant zones.

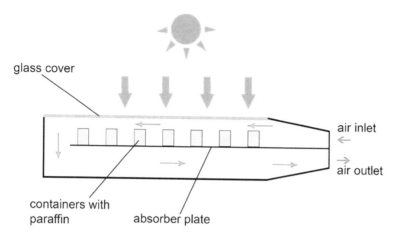

Fig. 8.11 Double-pass SAH with containers with paraffin [43]

Fig. 8.12 Boxes with PCM
in the drying chamber
[19]—top view

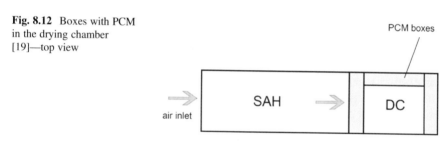

2.3.4 TES Inbuilt in the DC

An example of LHS inbuilt in the DC is reported in [19]. The PCM is placed in boxes in the DC in a fish dryer (Fig. 8.12). Three boxes of 3 kg paraffin are placed at the three sides of the DC.

A simple small-scale natural-convection solar dryer for 1 kg red chili [48] operated in the critical conditions of Nigerian climate, such as high humidity, very low average radiation intensity, and frequent rainfalls. It had a compact configuration (the SAH, 500x500x70 mm, and the DC, 1000x500 mm, are connected in a single unit) and good insulation of all structural parts with particle board. The heat storage was pellets of sodium sulfate decahydrate (Glauber's salt—$Na_2SO_4.10H_2O$) as PCM or sodium chloride (NaCl) as sensible heat storage placed in the SAH in charging mode and brought into the DC in discharging mode. A detailed energy and exergy analysis showed that the usage of $Na_2SO_4.10H_2O$ as PCM was the best option among the compared three systems, with $Na_2SO_4.10H_2O$, with NaCl and without thermal storage. The NaCl had no positive influence on the drying process. The specific energy consumption ranged from 3.34 to 5.92 kWh/kg, the solar dryer integrated with $Na_2SO_4.10H_2O$ having the lowest value.

3 Methods for MODELING and Simulation of Solar Dryers with Pcm

A variety of mathematical models are offered in literature for the purpose of comparison and design optimization of thermal storage systems. They predict the thermal behavior of the systems and calculate their efficiency. Two types of models are distinguished [1], based on the first law of thermodynamics (FLT) and the second law of thermodynamics (SLT). An overview of the first type models (FLT) [16] shows different analytical and numerical techniques to solve the energy equation at the moving solid–liquid interface (Stefan problem) of the PCM. The difficulty with the moving interface is avoided by the enthalpy method, proposed in [49]. It assumes a mushy zone between the two phases and introduces the total volumetric enthalpy as the sum of sensible heat and latent heat of the PCM. The energy conservation for the phase change process is expressed in terms of that enthalpy.

3.1 Energy Analysis

The energy analysis of a system with latent heat storage is based on the FLT. It calculates the energy efficiency, denoted also as first-law efficiency. The energy analysis presumes that energy efficiency η_I is defined as the ratio of the output energy E_{out} to the input energy E_{in}, part of which is lost as energy loss, E_{loss}

$$E_{in} = E_{out} + E_{loss} \qquad (8.1)$$

$$\eta_I = \frac{E_{out}}{E_{in}} = \frac{E_{in} - E_{loss}}{E_{in}} \qquad (8.2)$$

Energy efficiency equations, used by different authors for the constituent units in solar dryer systems with PCM storage, are listed below.

3.1.1 SAH Efficiency

The SAH efficiency is defined as a ratio of the heat flow rate extracted by the solar air heater to the solar radiation incident on the absorber surface of the solar collector.

$$\eta_{sah} = \frac{\dot{Q}_{SAH}}{A_{abs} I_{SAH}}, \qquad (8.3)$$

where $\dot{Q}_{SAH}$ is the heat flow rate, extracted by the SAH, which is calculated [38, 50] by:

$$\dot{Q}_{SAH} = \dot{m}_{SAH} c_{p,SAH} (T_{SAH,o} - T_{SAH,i}) \qquad (8.4)$$

3.1.2   Stored/Recovered Thermal Energy

The amount of energy stored inside the PCM during charging cycle Q_{ch} is expressed as:

$$Q_{ch} = m_{PCM} \left[c_{PCM,s} \left(T_{PCM,F} - T_{PCM,i_{ch}} \right) + L + c_{PCM,l} \left(T_{PCM,f_{ch}} - T_{PCM,F} \right) \right].$$
(8.5)

Similarly, the amount of energy recovered from the PCM during discharging cycle Q_{dis} is

$$Q_{dis} = m_{PCM} \left[c_{PCM,l} \left(T_{PCM,i_{dis}} - T_{PCM,F} \right) + L + + c_{PCM,s} \left(T_{PCM,F} - T_{PCM,f_{dis}} \right) \right].$$
(8.6)

3.1.3   Efficiency of the PCM Modules

The efficiency of the PCM modules is calculated by the ratio of the thermal energy extracted in discharging cycle to the thermal energy stored in charging cycle for a specific period of time [51].

$$\eta_{PCM} = \frac{Q_{dis}}{Q_{ch}}$$
(8.7)

3.1.4   Efficiency of the Solar Drying System Integrated with SAHs at Different Charging Modes

Natural Convection of Air through the SAHs without PCM Modules

The system efficiency is the ratio of the thermal energy used to evaporate the moisture from the sample to the global solar radiation incident on the absorber surface of the SAHs [35]:

$$\eta_{sys,1} = \frac{\int_0^{t_{end}} m_e \, h_{fg} \, dt}{\int_0^{t_{end}} I_{SAH} \, A_{abs} \, dt}.$$
(8.8)

Forced Convection of Air through the SAHs without PCM Modules

The system efficiency is the ratio of the thermal energy used to evaporate the moisture from the sample to the sum of the global solar radiation incident on the absorber surface and the electric energy consumption of the fan [35]:

$$\eta_{sys,2} = \frac{\int_0^{t_{end}} m_e \, h_{fg} \, dt}{\int_0^{t_{end}} I_{SAH} \, A_{abs} \, dt + \int_0^{t_{end}} P_{fan} \, dt}. \tag{8.9}$$

Forced Convection of Air through the SAHs with N PCM Modules

The system efficiency is the ratio of the thermal energy used to evaporate the moisture from the product plus the thermal energy stored in the PCM modules to the sum of the global solar radiation incident on the absorber surface and the electric energy consumption of the fan.

$$\eta_{sys,3} = \frac{\int_0^{t_{end}} m_e \, h_{fg} \, dt + \sum_o^n \int_0^{t_{end}} Q_{ch} \, dt}{\int_0^{t_{end}} I_{SAH} \, A_{abs} \, dt + \int_0^{t_{end}} P_{fan} \, dt}. \tag{8.10}$$

3.2 Exergy Analysis

The SLT models are introduced as a more correct approach to find the potential for improvement of the thermodynamic behavior of the heat storage [1]. According to them not energy is important but the thermodynamic availability of this energy. The loss of energy due to irreversibility of the process is taken into account. The SLT models formulate second-law efficiency based on entropy number [52] or exergy analysis. The benefit from exergy analysis is demonstrated by the comparison of exergy with energy given in Table 8.1 [53]. Based on literature, exergy is defined [54] as the maximum amount of work which can be produced by a system or a flow of matter till the system or the flow comes to equilibrium with a reference environment.

Second-law efficiency η_{II} can be expressed [55] as the ratio of exergy output Ex_{out} to the exergy input Ex_{in}:

$$\eta_{II} = \frac{Ex_{out}}{Ex_{in}} = \frac{Ex_{in} - Ex_{loss} - Ex_{des}}{Ex_{in}}. \tag{8.11}$$

Ex_{des} is the exergy destruction due to the irreversibility of the process, Ex_{loss} is the exergy lost to the environment.

Table 8.1 Comparison of energy and exergy [53]

Energy	Exergy
• is dependent on the parameters of matter or energy flow only, and independent of the environment parameters.	• is dependent both on the parameters of matter or energy flow and on the environment parameters.
• has the values different from zero (equal to mc² upon Einstein's equation).	• is equal to zero (in dead state by equilibrium with the environment).
• is governed by the FLT for all the processes.	• is governed by the FLT for reversible processes only (in irreversible processes it is destroyed partly or completely).
• is limited by the SLT for all processes (incl. Reversible ones).	• is not limited for reversible processes due to the SLT.
• is motion or ability to produce motion.	• is work or ability to produce work.
• is always conserved in a process, so can neither be destroyed nor produced.	• is always conserved in a reversible process, but is always consumed in an irreversible process.
• is a measure of quantity only.	• is a measure of quantity and quality due to entropy.

The exergy analysis in [35] of the units of a solar dryer with PCM uses the following equations taken from literature.

3.2.1   Exergy Analysis of the SAH

The exergy analysis employs the steady flow exergy equation [56] expressed as follows:

$$\dot{Ex} = \dot{m}_a \left\{ c_{pa}(T - T_r) - T_r \left[c_{pa} \ln\left(\frac{T}{T_r}\right) - R \ln\left(\frac{P}{P_r}\right) \right] \right\}. \tag{8.12}$$

The exergy efficiency of the SAH $\eta_{Ex,SAH}$ is expressed as the ratio between the exergy received by the working fluid (air) $\dot{Ex}_{re,air}$ the exergy inflow $\dot{Ex}_{in,SAH}$

$$\eta_{Ex,SAH} = \frac{\dot{Ex}_{re,air}}{\dot{Ex}_{in,SAH}}, \tag{8.13}$$

where the exergy inflow into the air heater is expressed as

$$\dot{Ex}_{in,SAH} = \left[1 - \frac{T_r}{T_{sun}}\right] \dot{Q}_{in}. \tag{8.14}$$

T_{sun} denotes the apparent sun temperature and it is assumed to be 4500 K, T_r is reference temperature of the ambient, $\dot{Q}_{in} = \alpha' \tau' IA_{SAH}$ is the energy input to the solar heater. $\dot{Ex}_{re,air}$ is expressed using the steady flow exergy equation (Eq. 8.12):

$$\dot{Ex}_{re,air} = \dot{m}_a c_{pa}\left[(T_{o,SAH} - T_{i,SAH}) - T_r \ln\left(\frac{T_{o,SAH}}{T_{i,SAH}}\right)\right]. \tag{8.15}$$

3.2.2 Exergy Analysis of TES

The exergy efficiency of TES is the ratio of the net exergy recovered from the energy storage during the discharging period Ex_{dis} to the net exergy input to the storage during the charging period Ex_{ch} [35].

$$\eta_{Ex,es} = \frac{Ex_{dis}}{Ex_{ch}}, \tag{8.16}$$

$$Ex_{ch} = \int_0^t \dot{m}_a c_{pa}\left[(T_{i,es} - T_{o,es}) - T_r \ln\left(\frac{T_{i,es}}{T_{0,es}}\right)\right] dt, \tag{8.17}$$

$$Ex_{dis} = \int_0^t \dot{m}_a c_{pa}\left[(T_{o,es} - T_{i,es}) - T_r \ln\left(\frac{T_{o,es}}{T_{i,es}}\right)\right] dt. \tag{8.18}$$

3.2.3 Exergy Analysis of the DC

The exergy efficiency of the DC, $\eta_{Ex,d}$, is defined as the ratio of the exergy outflow $\dot{Ex}_{od}$ to the exergy inflow $\dot{Ex}_{id}$ of the DC

$$\eta_{Ex,d} = \frac{\dot{Ex}_{od}}{\dot{Ex}_{id}}, \tag{8.19}$$

$$\dot{Ex}_{id} = \dot{m}_{da} c_{pa}\left[(T_{id} - T_r) - T_r \ln\left(\frac{T_{id}}{T_r}\right)\right], \tag{8.20}$$

$$\dot{Ex}_{od} = \dot{m}_{da} c_{pa}\left[(T_{od} - T_r) - T_r \ln\left(\frac{T_{od}}{T_r}\right)\right]. \tag{8.21}$$

3.3 Examples of Efficiency Evaluation of Solar Dryers with PCM

The efficiencies of the units of a solar dryer with thermal energy storage using paraffin wax as PCM were evaluated in [30]. A mathematical model was developed and validated against experimental data. It predicted important operation characteristics like heat transfer and drying parameters, as well as ambient parameters and solar radiation. The efficiency of the PCM-based accumulator was calculated as the ratio of the heat absorbed by the air in the accumulator $Q_{abs,f}$ to the heat stored by the paraffin wax $Q_{abs,w}$

$$\eta_{SA} = \frac{Q_{abs,f}}{Q_{abs,w}} 100\%. \tag{8.22}$$

The efficiency was obtained in the range of 45.5–63% for different mass flow rates of the drying air. As a result of the numerical simulation, it was found that after the sunshine hours and at zero solar radiation the drying air was heated for 4 to 8 hours to an average drying air temperature 10–27.7 °C above the ambient temperature depending on the air flow rate.

The article [35] presents performance studies of a forced-convection solar dryer of chili, integrated with a shell-and-tube TES unit. The performance of each component of the drying system was evaluated in terms of energy and exergy efficiency. The average instantaneous heat input and heat recovered during the charging and discharging processes of the energy storage were in the range of 105–130 W and 89–116 W, respectively. The net heat input and heat recovered varied from 2.5 MJ to 3.2 MJ and from 1.2 MJ to 1.5 MJ, respectively. The average energy efficiency (the percentage of the energy recovered) (Eq.8.7) was in the range of 43.6–49.8%. No heat energy was retrieved from the storage after 18:00 h, as the air coming out of the storage at a temperature below 36 °C was not much effective for drying. The drying time was reduced by 55% of the drying time in open sun drying. The TES provided an extended drying time of 2 h after sunset and the overall efficiency of the drying system was 10.8%. The net exergy input and exergy recovered were found to be in the range of 0.2–0.3 MJ and 0.04–0.05 MJ, respectively. The overall exergy efficiency ranged from 18.3% to 20.5%.

Pebble stones and paraffin wax (as TES) were used in drying of lemon slices [3]. Considering that the drying time increased with increasing the product amount, it was concluded that maximum 11.3 kg of lemon slices could be dried with each of the energy storage systems. As a result of the experimental studies, the average energy efficiencies were obtained as 68.2% and 68.55% for pebble stones and paraffin, respectively. When the systems were considered economically, it was observed that pebble stones TES had a 10.47% lower initial investment cost compared to paraffin. Pebble stone unit price used in the system was $ 0.0725/kg. The unit price of paraffin wax used in the PCM system was $ 2.483/kg. Moreover, paraffin wax loses its ability to store energy since its properties deteriorate after a certain period of time (approximately 4 years) and it should be replaced.

Energy-based performance and exergy-based performance of a natural-convection solar dryer integrated with sodium sulfate decahydrate and sodium chloride as thermal storage are presented [48]. Evaluation of the thermal storage potential of the two materials and a control experiment (without thermal storage) with a focus on energy consumption and exergy sustainability indicators was performed. It showed that the moisture content of red chili is reduced at most for $Na_2SO_4 \cdot 10H_2O$. The overall drying efficiency and energy consumption of the three cases varied from 10.61 to 18.79% and 7.54–12.98 MJ, respectively. The specific energy consumption ranged from 3.34 to 5.92 kWh/kg with a solar dryer integrated with $Na_2SO_4 \cdot 10H_2O$ having the least value. The exergy efficiency for the three cases laid between 66.79% and 96.09% with average values of 82.3%, 82.69%, and 82.65%, respectively.

The effects of a v-corrugated absorber plate and integration of PCM in the SAH were studied [44]. It was concluded that when using PCM, the daily energy efficiency of the v-corrugated solar heater was 62%, 52%, and 27% compared with 50%, 43.2%, and 22.2% for the same heater but, without using PCM at air mass flow rate 0.062, 0.028, and 0.009 kg/s, respectively. The daily efficiency for the flat plate with PCM was 47%, 35.2%, and 18.2% compared with 40.7%, 32%, and 14.4% for the same heater without PCM at air mass flow rate 0.062, 0.028, and 0.009 kg/s, respectively. It was found that 4 cm PCM thickness provided 1 hour more discharging time and 2 degrees more temperature difference than 2 cm PCM.

An indirect forced-convection solar dryer integrated with different sensible heat storage materials for drying chili was developed and evaluated [57]. The inclusion of the heat storage material increased the drying time by about 4 h per day. The chili was dried from initial moisture content of 72.8% to the final moisture content of about 9.2% and 9.7% (wet basis) at the bottom and the top tray, respectively. The calculated thermal efficiency of the solar dryer was about 21%. The specific moisture extraction rate was about 0.87 kg/kWh.

3.4 Mathematical Modeling of the Thermal Behavior of a Solar Dryer with PCM

3.4.1 CFD with Enthalpy-Porosity Model

Computational fluid dynamics (CFD) is a useful tool in developing of heat storage. CFD is employed in investigation of air flow pattern, temperature distribution, and humidity distribution inside the units of the solar dryer. It reveals the picture of melting and solidification of PCM and the phases' distribution in space and time. The typical numerical packages for CFD modeling of solar dryers with thermal storage [58] include: COMSOL Multiphysics, ANSYS CFX and Fluent, FORTRAN and OpenFOAM. The CFD simulations can substitute the physical drying experiments. After validation by experimental data, the CFD model is applied for predictions in a variety of drying conditions and designs for finding the optimal ones.

CFD Simulation of a Single Container with PCM

ANSYS Fluent was used in [59] for modeling of solidification and melting of paraffin macro-encapsulated in a stainless steel container. Fluent software adopts the enthalpy-porosity approach [49], which avoids tracking explicitly the melt interface. The fraction of the cell volume which is in liquid form is computed at each iteration, based on the enthalpy balance. The mushy zone, where the liquid fraction lies between 0 and 1, is modeled as "pseudo" porous medium in which the porosity in each cell is set equal to the liquid fraction and decreases from 1 to 0 as the material solidifies, and hence the velocity drops to zero for solid state.

The geometry in [59] was taken from an experimental TES constructed at the Technical University (TU) Sofia, Plovdiv Brunch (Fig. 8.13). The containers with PCM with external dimensions 40x80x950 mm were placed vertically in a tank with water as a HTF heated by solar energy. The wall thickness was 1.5 mm. The transient thermal behavior of a single container with paraffin was simulated. Results were obtained at various temperatures of the external surfaces of the container, $T_{wall} = 15°C$ for discharging mode, $T_{wall} = 65°C$ and $80°C$ for charging mode. The simulation revealed the flow pattern and the distribution of the temperature and the PCM phases at these conditions.

The 3D computational domain was discretized by a mesh of 26,000 tetrahedral elements. The density difference between the solid and the liquid phase was assumed negligible. The natural convection was expected to play an important role especially in a tall vertical tube like the unit under consideration, Fig. 8.13. A laminar flow of incompressible Newtonian fluid was assumed for the liquid phase. The buoyant forces were modeled by the Boussinesq approximation.

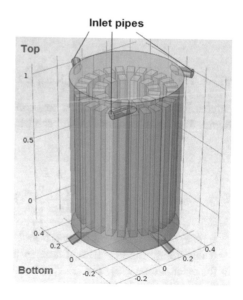

Fig. 8.13 Containers with paraffin TES at TU Sofia, Plovdiv Branch [60]

The governing equations for the liquid PCM include the following:
Continuity equation:

$$\frac{\partial \rho}{\partial t} + \nabla(\rho v) = 0. \tag{8.23}$$

Momentum equation:

$$\frac{\partial}{\partial t}(\rho v) + \nabla(\rho vv) = -\nabla p + \nabla \tau + \rho g + S, \tag{8.24}$$

where v is the velocity vector, p is the static pressure, τ is the stress tensor, and ρg and S are the vectors of gravitational and external body forces.

The momentum sink due to the reduced porosity in the mushy zone takes the following form:

$$S = \frac{(1-\beta)^2}{(\beta^3 + \varepsilon)} A_{mush} v, \tag{8.25}$$

where β is the liquid volume fraction, ε is a small number (0.0001) to prevent division by zero, $A_{mush} = 10^5$ is the mushy zone constant.

Energy equation:

$$\frac{\partial}{\partial t}(\rho H) + \nabla(\rho v H) = \nabla(k\nabla T). \tag{8.26}$$

The enthalpy H consists of sensible enthalpy h and latent heat ΔH:

$$H = h + \Delta H, \tag{8.27}$$

where $h = h_r + \int_{T_r}^{T} c_p dT$, $\Delta H = \beta L$; c_p is specific heat capacity, J/(kgK); k—thermal conductivity, W/(mK). The latent heat content can vary between 0 for solid and L for liquid. The liquid volume fraction can be defined as:

$$\begin{aligned} \beta &= 0 & \text{if } T < T_s \\ \beta &= 1 & \text{if } T > T_l \\ \beta &= \frac{T - T_s}{T_l - T_s} & \text{if } T_s < T < T_l, \end{aligned} \tag{8.28}$$

where T_s and T_l are solidus and liquidus temperatures, respectively.

Table 8.2 gives the values of the thermo-physical properties of the paraffin used in the calculations [59]. The accepted constants are based on data for E53 paraffin from [61]. The model does not take into account the experimental observations [61]

Table 8.2 Thermo-physical properties, based on [61]

Solidus temperature, T_s, °C	52
Liquidus temperature, T_l, °C	59
Density (solid), kg/m³	920
Dynamic viscosity, kg/ms	0.003
Specific heat (solid), J/kgK	1550
Thermal conductivity (solid), W/mK	0.34
Latent heat of fusion, J/kg	194,110
Thermal expansion coefficient, 1/K	0.0001

that the paraffin properties depend on the temperature and the physical state. Moreover, E53 paraffin undergoes two-step phase transition, the lower temperature transition is solid–solid (order-disorder) transition, while the higher temperature transition represents the solid–liquid phase change process. The model assumes that in heating and cooling only solid–liquid transition occurs over the higher temperature range of 52–59°C, yet the accepted value of the latent heat is the cumulative value of the two steps.

The eqs. (23–26) were solved [59] using ANSYS Fluent 13.0, based on finite volume technique. The following boundary conditions (BC) were assumed: no-slip conditions at the container's wall and constant temperature at all external surfaces $T = T_{wall}$. In charging mode, the initial conditions were solid PCM with temperature $T_{in} = 51°C$, one degree lower than the solidus temperature, given in Table 8.2; in discharging mode, $T_{in} = 60°C$, one degree higher than the liquidus temperature, Table 8.2. All the surfaces were at constant temperature.

The simulation results [59] in Fig. 8.14 show that after 20 min charging the liquid fraction is 88%, and due to the buoyant forces the paraffin is stratified, liquid at the top, solid at the bottom. Due to natural convection, there is an upward flow along the side walls of the container, and a downward flow in the central zone. Therefore, the convective heat transfer, often neglected in the simulations, affects substantially the picture of the melting process. However, as expected, it has negligible effect on the solidification in the container. In some cases, conductive heat transfer can be dominant in the initial stage of the phase change process and natural convection becomes dominant with the progress of the melting process [62].

It was found [21] that the natural convection in the PCM was dependent on the HTF flow direction during melting. The natural convection was dominant for upward flow of HTF and negligible for downward flow. The solidification process was dominated by conduction and independent of the HTF direction.

The results [59] in Fig. 8.15 illustrate the effect of the external temperatures on the duration of the phase change process. A simulation was performed at a constant heat flux through the wall with a convective heat transfer coefficient $\alpha = 3000$ W/(m²K), calculated by the relation for laminar boundary layer of water flow on a flat surface [63] at a bulk fluid temperature $T_{fl} = 65°C$. The results in Fig. 8.15 in that case coincide with those calculated at a constant wall temperature $T_{wall} = 65°C$, due

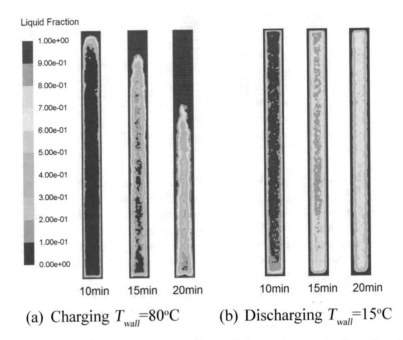

Liquid Fraction

(a) Charging T_{wall}=80°C   (b) Discharging T_{wall}=15°C

Fig. 8.14 Contours of liquid fraction in the middle-width plane parallel to the 80 mm face, (**a**) in charging at $T_{wall} = 80°C$, (**b**) in discharging at $T_{wall} = 15°C$ [59]

to the low thermal resistance at the fluid-solid interface, therefore the negligibly small temperature difference between the fluid bulk and the solid surface of the container.

Similar set of ANSYS Fluent solution methods and settings was used in [62]. The computational model was verified by comparing the simulation results with experimental data for melting of paraffin wax n-octadecane (28.2°C melting temperature) encapsulated in a spherical container of glass with inside diameter of 101.66 mm. It was found that the prediction of the overall melting process was satisfactory.

CFD Simulation of a PCM-Based TES Unit

Transient simulation of the TES unit (Fig. 8.13) was conducted in a 3D domain by means of COMSOL multiphysics (based on a finite element method) [60]. It was assumed that the volume of the PCM did not change during phase transition. In order to simulate the dynamic and thermal behavior of the HTF flowing inside the TES, the continuity equation, the Navier–Stokes equations, and the energy equation for laminar flow were solved.

The energy equation for the PCM including latent heat transfer during phase change is given by:

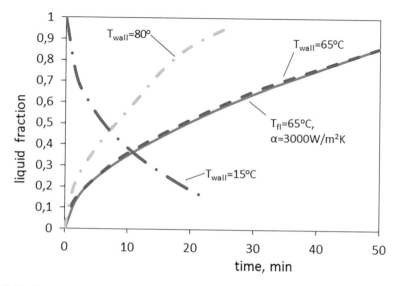

Fig. 8.15 Computed liquid volume fraction in the total PCM volume (0 when the total volume is in solid state and 1 when the total volume is in liquid state) versus time in the phase change process [59]

$$\rho c_p \frac{\partial T}{\partial t} + \rho c_p \mathbf{v} \cdot \nabla T = \nabla \cdot (k \nabla T) \qquad (8.29)$$

The equivalent thermal conductivity, specific heat and density were expressed in [60] as a mass weighted average of the total liquid and solid mass of the PCM [64]. The expressions [64] assume smooth transition over the temperature range. The latent heat of fusion is taken into consideration in the equivalent specific heat.

In the PCM containers only conduction heat transfer was considered in [60] without buoyancy effects or subcooling during the phase change processes. For the fluid flow, the BC at the solid surfaces (the inner wall of the storage tank and the surface of the PCM containers) were considered as no-slip BC with constant wall temperature. It was assumed that the storage tank was perfectly insulated. The containers were highly conductive layers (stainless steel) transferring heat between the HTF and the PCM. Outlet BCs for the velocity were set up in terms of pressure with suppressed backflow. The outlet temperature BC was Neumann type. The simulation [60] studied the effect of the number of inlet and outlet pipes (1–3). The initial temperature of the PCM and HTF in the storage tank was 20°C in charging and 80°C in discharging. The inlet velocity was taken depending on the number of the inlet pipes. The storage was charged with a net heat rate of 1 kW through the inlet pipes. In the discharging process 1 kW of thermal energy left from the outlet pipes, and the inlet flow was maintained at 20°C. Paraffin E46 [61] was considered as PCM. The HTF velocity at the inlet was 0.1 m/s.

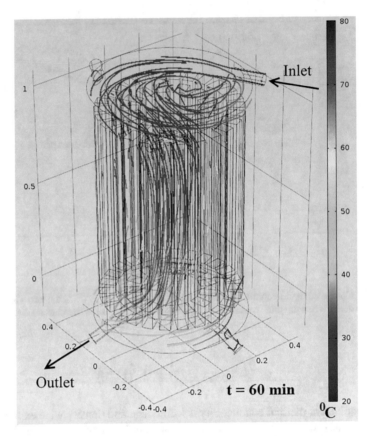

Fig. 8.16 Charging process with a single inlet and a single outlet pipe: HTF streamlines with thermal effect at 60 min [60]

Simulation results [60] are presented in Fig. 8.16 for the case of one inlet and one outlet pipe in charging mode. The HTF enters the tank, circulates over the PCM containers and the most of the fluid flows downward through the central zone of the tank, thus charging the inner PCM containers. It can be seen from the figure that after 60 min the temperature distribution inside the storage is uneven and the PCM containers are not fully charged.

From the obtained temperature and velocity distribution it was concluded [60] that two inlet and two outlet pipes were enough to charge or discharge the TES in three hours. The PCM containers in the inner zone were charged first. The outer containers needed more time for charging because of their location, close to the tank walls, which hindered the fluid flow and the heat transfer process.

A low-temperature latent heat TES device was studied [65] for drying of agricultural products in an indirect type solar dryer. A 2D geometry was created. The air flowed into an inner copper tube and the PCM was in an outer plastic tube. Transient simulations were conducted by ANSYS Fluent 2015 to capture the distribution of velocity, temperature, and liquid/solid fractions. The CFD simulations [65] were run from 8.00 am to 10.00 pm. The inlet air temperature was gradually increased and reached a maximum value of 349 K at 2.00 pm. The model helped to analyze the potential of PCM to store excess solar energy. It estimated the temperature distribution in the radial direction of the TES device. A 2D numerical study was conducted [66] for performance improvement of the TES from [65] by fins on the outer surface of the copper tube. Two cases were considered: Case-I without fins and Case-II with fins. The TES dimensions of both cases were the same. Fin dimensions were 0.5 mm tip diameter, 5 mm length, 14.86 mm fin spacing. Paraffin wax was used as a PCM material for both cases. CFD simulations were carried out for four air velocity conditions (1 m/s, 2 m/s, 3 m/s, 4 m/s). It was found that both cases worked fine with an air velocity of 1 m/s compared to higher air flow velocities. Higher air flow velocity did not affect much the DC temperature compared to lower velocity. Also, lower air flow velocity maintained the uniformity of drying over a longer period of time. At the same time, higher air flow velocity carried more moisture from the food product, so the drying time could be reduced. Melting fraction was higher in Case-II compared to Case-I as the fins transferred more heat to the PCM. For all the velocities considered in this work, the heat gained by the air was higher for Case-II compared to Case-I. The maximum heat gained by the air for Case-II was 55.2% more in comparison to Case-I at an air velocity of 1 m/s.

CFD Simulation of the DC

The proper conditions in the DC are the main purpose of modeling and simulation of the dryer system. A mathematical modeling of a solar agricultural dryer with a back-up biomass burner and sensible TES is presented in [67], compared to experimental observations. The solar dryer system consists of a solar air heater, drying chamber, and a back-up heater burning biomass. The biomass burner is surrounded by the TES space filled with bricks. The bottom plate of the drying chamber is placed directly on top of the TES unit. A 40-mm free space between the internal and external walls of the DC forms a "jacket" around three sides of the chamber, which allows the exhausted gas that passes through the thermal storage to flow, before it is released to the ambient area, through the chimney and to keep the DC warm. The bricks used for heat storage are arranged in a manner in which the exhaust gas and smoke from combustion pass between them, before venting out to the atmosphere, in order to maximize the capture of heat from the exhausted gas.

A FLT model was suggested [67] for thermal analysis of the system units and calculation of the mass exchange between the product and the drying air. Additionally, a CFD simulation (by FloVent program) was used to predict the temperature distribution and the air flow pattern in the drying chamber for the situation when the back-up heater was in operation.

The governing equations for the turbulent natural convection air flow are the Reynolds-Averaged Navier–Stokes equations given by:

Equation of continuity:

$$\frac{\partial}{\partial x_i}(\rho \overline{u_i}) = 0 \tag{8.30}$$

Navier–Stokes equation:

$$\frac{\partial}{\partial x_i}(\rho \overline{u_j} \overline{u_i}) = -\frac{\partial \overline{P}}{\partial x_i} + \frac{\partial}{\partial x_j}\left[\mu\left(\frac{\partial \overline{u_i}}{\partial x_j} + \frac{\partial \overline{u_j}}{\partial x_i}\right) - \rho \overline{u_i' u_j'}\right] - \rho g_i \gamma (\overline{T} - \overline{T}_r), \tag{8.31}$$

Energy equation:

$$\frac{\partial}{\partial x_i}(\rho \overline{u_i T}) = \frac{\partial}{\partial x_i}\left[\frac{\mu}{Pr}\frac{\partial T}{\partial x_i} - \rho \overline{u_j' T'}\right] \tag{8.32}$$

where $\overline{u_i}$ indicates time-averaged velocity component, $\overline{u_i'}$ is velocity fluctuation, x_i is the coordinate axis, g_i is the gravitational acceleration vector and γ is the thermal expansion coefficient, μ is the dynamic viscosity, Pr is Prandtl number. The Boussinesq approximation is employed in the last term of Eq. (8.31), where T_r is the reference temperature, $\overline{T}$ is the mean temperature, and T' is the temperature fluctuation. A k-ε model is used to correlate the Reynolds terms $\rho \overline{u_i' u_j'}$ and $\rho \overline{u_j' T'}$ to the mean flow field.

The obtained temperature distribution in the DC is presented in Fig. 8.17. It was concluded in [67] that the average air temperature in the DC of 56 °C was suitable for drying of agricultural products.

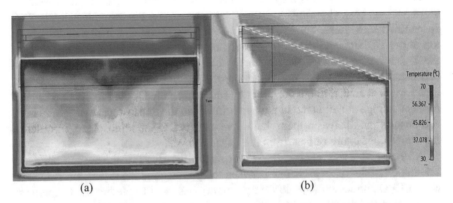

Fig. 8.17 CFD simulation of temperature distribution of air in the DC [67] (CC-BY license)

4 Conclusions

Modeling and simulation of the TES in a solar dryer system is a tool for design improvement and increasing the efficiency of the system. The challenge is that the models should take into consideration various factors affecting the rate of drying, such as solar radiation, ambient temperature, air flow velocity, relative humidity, initial moisture content, wind speed, type of goods, absorptivity, and mass of product per unit exposed area. The prediction of the conditions for heat transfer in the units of the solar dryer with paraffin (containers' materials and surface area, insulation, desiccants, air flow rate and velocity, arrangement of the grids for the drying material, etc.) gives the possibility to improve the performance characteristics of a successful simplified small-size design.

The present overview identifies the main requirements in the design of a solar dryer with PCM and the most appropriate modeling approaches for performance improvement, as follows:

- Stable temperature of the drying air at least 10–25 °C above the ambient temperature. This is necessary to avoid the product reabsorbing the moisture during night, when the air temperature drops and its humidity increases.
- Choice of proper PCM according to the drying regime; the minimum melting temperature of the PCM should be 5–10 °C higher than the desired temperature of the HTF.
- Minimal heat loss by effective thermal insulation.
- Intensive heat transfer (measures for reduction of air convective resistance and enhancement of thermal conductivity of PCM).
- Metal lamellae structures are a good solution for increasing paraffin thermal conductivity more than 2 times.
- Macro-encapsulation is recommended for providing large heat transfer area, reduction of the paraffin reactivity towards environment and controlling the volume changes as phase change occurs. Increasing the height/width ratio of the container for the same volume decreases the time for the melting process due to the stronger buoyancy effect. Usually the material of the shell is plastic or metal (copper, aluminum, steel) when higher heat transfer rates are desirable.
- Metal fins enhance significantly the heat transfer rate in and out of the element containing the PCM.
- The thermal storage should not create additional pressure drop, especially in natural convection units. It can also serve as an obstacle to the air flow, which increases the heat transfer area and enhances the air flow conditions by creating turbulence and eliminating stagnant zones.
- The second-law models are a more informative approach to find the potential for improvement of the thermodynamic behavior of the thermal energy accumulator, since they evaluate the thermodynamic availability of energy.
- CFD simulation of the TES is useful for analyzing the potential of PCM to store excess solar energy during sunshine hours and to release thermal energy at night. It enables to compare different device configurations by revealing the picture of temperature, humidity, and velocity distribution.

Acknowledgments This work is supported by the National Science Fund, Bulgaria, Contract No KP-06-INDIA/11/02.09.2019 and the Department of Science and Technology, India (DST/INT/P-04/2019).

References

1. P. Verma, K. Varun, S.K. Singal, Review of mathematical modeling on latent heat thermal energy storage systems using phase-change material. Renew. Sust. Energ. Rev. **12**, 999–1031 (2008)
2. J.L. Butler, J.M. Troeger, *Drying Peanuts Using Solar Energy Stored in a Rockbed. Agricultural Energy, Vol. 1 Solar Energy, Selected Papers and Abstracts* (ASAE Publication, St. Joseph, MI, 1980)
3. H. Atalay, Assessment of energy and cost analysis of packed bed and phase change material thermal energy storage system for the solar assisted drying process. Sol. Energy **198**, 124–138 (2020)
4. G. Alva, Y. Lin, G. Fang, An overview of thermal energy storage systems. Energy **144**, 341–378 (2018)
5. V. Shanmugam, E. Natarajan, Experimental study of regenerative desiccant integrated solar dryer with and without reflective mirror. Appl. Therm. Eng. **27**(8–9), 1543–1551 (2007)
6. A. Bonaparte, Z. Alikhani, C.A. Madramootoo, V. Raghavan, Some quality characteristics of solar-dried cocoa beans in St Lucia. J. Sci. Food Agric. **76**, 553–558 (1998)
7. V.K. Sharma, A. Colangelo, G. Spagna, Experimental investigation of different solar dryers suitable for fruit and vegetable drying. Renew. Energy **6**(4), 413–424 (1995)
8. C.L. Hii, C.L. Law, A. Rahman, S. Jinap, Y.B. Che Man, Quality comparison of cocoa beans dried using solar and sun drying with perforated and non-perforated drying platform. Proceedings of the 5th Asia-Pacific Drying Conference **12**, 546–552 (2007)
9. G.A. Lane, *Solar Heat Storage: Latent Heat Materials, Vol. I: Background and Scientific Principles* (CRC, Boca Raton, Florida, 1983)
10. S. Aboul-Enein, A.A. El-Sebaii, M.R.I. Ramadan, H.G. El-Gohary, Parametric study of a solar air heater with and without thermal storage for solar drying applications. Renew. Energy **21**(3–4), 505–522 (2000)
11. A.O. Fagunwa, O. A. Koya, and M.O. Faborode. Development of an intermittent solar dryer for cocoa beans. Agricultural Engineering International, Manuscript number 1292, vol. XI, July 2009, [E journal] Available: CIGR E journal
12. S.F. Dina, H. Ambarita, F.H. Napitupulu, H. Kawai, Study on effectiveness of continuous solar dryer integrated with desiccant thermal storage for drying cocoa beans. Case Studies in Thermal Engineering **5**, 32–40 (2015)
13. A. Sharma, C.R. Chen, N.V. Lan, Solar-energy drying systems: A review. Renew. Sust. Energ. Rev. **13**(6–7), 1185–1210 (2009)
14. A.K. Bhardwaj, R. Chauhan, R. Kumar, M. Sethi, A. Rana, Experimental investigation of an indirect solar dryer integrated with phase change material for drying valeriana jatamansi (medicinal herb). Case studies in thermal engineering **10**, 302–314 (2017)
15. S.M. Shalaby, M.A. Bek, A.A. El-Sebaii, Solar dryers with PCM as energy storage medium: A review. Renew. Sust. Energ. Rev. **33**, 110–116 (2014)
16. A. Sharma, V.V. Tyagi, C.R. Chen, D. Buddhi, Review on thermal energy storage with phase change materials and applications. Renew. Sust. Energ. Rev. **13**(2), 318–345 (2009)
17. L.M. Bal, S. Satya, S.N. Naik, Solar dryer with thermal energy storage systems for drying agricultural food products: A review. Renew. Sust. Energ. Rev. **14**(8), 2298–2314 (2010)
18. O.A. Babar, V.K. Arora, P.K. Nema, Selection of phase change material for solar thermal storage application: A comparative study. J. Braz. Soc. Mech. Sci. Eng. **41**(9), 355 (2019)

19. V.M. Swami, A.T. Autee, T.R. Anil, Experimental analysis of solar fish dryer using phase change material. Journal of Energy Storage **20**, 310–315 (2018)
20. R. Singh, S. Sadeghi, B. Shabani, Thermal conductivity enhancement of phase change materials for low-temperature thermal energy storage applications. Energies **12**(1), 75 (2019)
21. H. Ettouney, H. El-Dessouky, E. Al-Kandari, Heat transfer characteristics during melting and solidification of phase change energy storage process. Ind. Eng. Chem. Res. **43**(17), 5350–5357 (2004)
22. A. Khyad, H. Samrani, M.N. Bargach, R. Tadili, Energy storage with PCMs: Experimental analysis of paraffin's phase change phenomenon & improvement of its properties. J. Mater. Environ. Sci **7**(7), 2551–2560 (2016)
23. B. Cárdenas, N. León, High temperature latent heat thermal energy storage: Phase change materials, design considerations and performance enhancement techniques. Renew. Sust. Energ. Rev. **27**, 724–737 (2013)
24. L. Liu, D. Su, Y. Tang, G. Fang, Thermal conductivity enhancement of phase change materials for thermal energy storage: A review. Renew. Sust. Energ. Rev. **62**, 305–317 (2016)
25. S. Wu, T. Yan, Z. Kuai, W. Pan, Thermal conductivity enhancement on phase change materials for thermal energy storage: A review. Energy Storage Materials **25**, 251–295 (2020)
26. M.M. Farid, A.M. Khudhair, S.A.K. Razack, S. Al-Hallaj, A review on phase change energy storage: Materials and applications. Energy Convers. Manag. **45**(9–10), 1597–1615 (2004)
27. N.S. Dhaidan, J.M. Khodadadi, Melting and convection of phase change materials in different shape containers: A review. Renew. Sust. Energ. Rev. **43**, 449–477 (2015)
28. B. Zivkovic, I. Fujii, An analysis of isothermal phase change of phase change material within rectangular and cylindrical containers. Sol. Energy **70**(1), 51–61 (2001)
29. J. Wei, Y. Kawaguchi, S. Hirano, H. Takeuchi, Study on a PCM heat storage system for rapid heat supply. Appl. Therm. Eng. **25**(17–18), 2903–2920 (2005)
30. J. Vásquez, A. Reyes, N. Pailahueque, Modeling, simulation and experimental validation of a solar dryer for agro-products with thermal energy storage system. Renew. Energy **139**, 1375–1390 (2019)
31. S. Devahastin, S. Pitaksuriyarat, Use f latent heat storage to conserve energy during drying and its effect on drying kinetics of a food product. Appl. Therm. Eng. **26**, 1705–1713 (2006)
32. Z. Khan, Z.A. Khan, An experimental investigation of discharge/solidification cycle of paraffin in novel shell and tube with longitudinal fins based latent heat storage system. Energy Convers. Manag. **154**, 157–167 (2017)
33. A. Tiwari, A review on solar drying of agricultural produce. J. Food Process. Technol. **7**(9), 1–12 (2016)
34. D.V.N. Lakshmi, P. Muthukumar, A. Layek, P.K. Nayak, Drying kinetics and quality analysis of black turmeric (Curcuma caesia) drying in a mixed mode forced convection solar dryer integrated with thermal energy storage. Renew. Energy **120**, 23–34 (2018)
35. D.K. Rabha, P. Muthukumar, Performance studies on a forced convection solar dryer integrated with a paraffin wax-based latent heat storage system. Sol. Energy **149**, 214–226 (2017)
36. V.V. Bhagwat, S.P. Salve, S. Debnath, Experimental analysis of a solar dehydration with phase changing material. AIP Conference Proceedings **1**, 020003 (1998)
37. S. Kumar, V.S.K. Kumar, Charging-discharging characteristics of macro encapsulated phase change materials in an active thermal energy storage system for a solar drying kiln. Therm. Sci. **21**, 2525–2532 (2017)
38. A. El Khadraoui, S. Bouadila, S. Kooli, A. Farhat, A. Guizani, Thermal behavior of indirect solar dryer: Nocturnal usage of solar air collector with PCM. J. Clean. Prod. **148**, 37–48 (2017)
39. H. Esen, Experimental energy and exergy analysis of a double-flow solar air heater. Build. Environ. **43**, 1046–1054 (2008)
40. H. Benli, Experimentally derived efficiency and exergy analysis of a new solar air heater having different surface shapes. Renew. Energy **50**, 58–67 (2013)

41. S. Karthikeyan, G. Ravikumar Solomon, V. Kumaresan, R. Velraj, Parametric studies on packed bed storage unit filled with PCM encapsulated spherical containers for low temperature solar air heating application. Energy Convers. Manag. **78**, 74–80 (2014)
42. A. Reyes, A. Mahn, F. Vásquez, Mushrooms dehydration in a hybrid-solar dryer, using a phase change material. Energy Convers. Manag. **83**, 241–248 (2014)
43. A.K. Raj, M. Srinivas, S. Jayaraj, A cost-effective method to improve the performance of solar air heaters using discrete macro-encapsulated PCM capsules for drying applications. Appl. Therm. Eng. **146**, 910–920 (2019)
44. A.E. Kabeel, A. Khalil, S.M. Shalaby, M.E. Zayed, Experimental investigation of thermal performance of flat and v-corrugated plate solar air heaters with and without PCM as thermal energy storage. Energy Convers. Manag. **113**, 264–272 (2016)
45. T. Alam, R.P. Saini, J.S. Saini, Experimental investigation on heat transfer enhancement due to V-shaped perforated blocks in a rectangular duct of solar air heater. Energy Convers. Manag. **81**, 374–383 (2014)
46. A. Ghiami, S. Ghiami, Comparative study based on energy and exergy analyses of a baffled solar air heater with latent storage collector. Appl. Therm. Eng. **133**, 797–808 (2018)
47. R. Bakari, Heat transfer optimization in air flat plate solar collectors integrated with baffles. Journal of Power and Energy Engineering **6**(1), 70–84 (2018)
48. M.C. Ndukwu, L. Bennamoun, F.I. Abam, A.B. Eke, D. Ukoha, Energy and exergy analysis of a solar dryer integrated with sodium sulfate decahydrate and sodium chloride as thermal storage medium. Renew. Energy **113**, 1182–1192 (2017)
49. V.R. Voller, M. Cross, N.C. Markatos, An enthalpy method for convection/diffusion phase change. Int. J. Numer. Meth. Eng. **24**, 271–284 (1987)
50. S.M. Shalaby, M.A. Bek, Experiment investigation of a novel indirect solar dryer implementing PCM as energy storage medium. Energy Convers. Manag. **83**, 1–8 (2014)
51. A.G. Georgiev, R. Popov, and E.T. Toshkov. Investigation of a hybrid system with ground source heat pump and solar collectors: Charging of thermal storages and space heating. Renew. Energy, vol. 147, part 2, pp. 2774–2790, 2020
52. A. Bejan, Two thermodynamic optima in the design of sensible heat units for energy storage. J. Heat Transf. **100**(4), 708–712 (1978)
53. I. Dincer, Y.A. Cengel, Energy, entropy and exergy concepts and their roles in thermal engineering. Entropy **3**, 116–149 (2001)
54. K. Taheri, R. Gadowa, A. Killinger, Exergy analysis as a developed concept of energy efficiency optimized processes: The case of thermal spray processes. Procedia CIRP **17**, 511–516 (2014)
55. D. Kumar, P. Mahantaa, P. Kalita, Energy and exergy analysis of a natural convection dryer with and without sensible heat storage medium. Journal of Energy Storage **29**, 101481 (2020)
56. P.K. Nag, *Basic and Applied Thermodynamics* (Tata McGraw-Hill, 2006)
57. M. Mohanraj, P. Chandrasekar, Performance of a forced convection solar dryer integrated with gravel as heat storage material for chili drying. J. Eng. Sci. Technol. **4**(3), 305–314 (2009)
58. K. Kant, A. Shukla, A. Sharma, A. Kumar, A. Jain, Thermal energy storage based solar drying systems: A review. Innovative Food Sci. Emerg. Technol. **34**, 86–99 (2016)
59. D. B. Dzhonova-Atansova, A. G. Georgiev, and R. K. Popov. Numerical study of heat transfer in macro-encapsulated phase change material for thermal energy storage. Bulg. Chem. Commun., vol. 48, Spec. Iss. E, pp.189–194, 2016
60. A. Seitov, B. Akhmetov, A. G. Georgiev, A. Kaltayev, R. K. Popov, D. B. Dzhonova-Atanasova, and M. S. Tungatarova. Numerical simulation of thermal energy storage based on phase change materials. Bulg. Chem. Commun., vol. 48, Spec. Iss. E, pp. 181–188, 2016
61. E.M. Anghel, A. Georgiev, S. Petrescu, R. Popov, M. Constantinescu, Thermo-physical characterization of some paraffins used as phase change materials for thermal energy storage. J. Therm. Anal. Calorim. **117**, 557–566 (2014)

62. F.L. Tan, S.F. Hosseinizadeh, J.M. Khodadadi, L. Fan, Experimental and computational study of constrained melting of phase change materials (PCM) inside a spherical capsule. Int. J. Heat Mass Transf. **52**, 3464–3472 (2009)
63. J. H. Leinhard IV, J. H. Leinhard V, A Heat Transfer Textbook, 3rd ed. Cambridge MA, Phlogiston Press, 2008
64. COMSOL, *Multiphysics Model Library* (COMSOL Inc., 2012)
65. S. Yadav, A.B. Lingayat, V.P. Chandramohan, V.R.K. Raju, Numerical analysis on thermal energy storage device to improve the drying time of indirect type solar dryer, in *Heat and Mass Transfer*, (Springer, Cham, 2018). https://doi.org/10.1007/s00231-018-2390-7
66. S. Yadav, V.P. Chandramohan, Performance comparison of thermal energy storage system for indirect solar dryer with and without finned copper tube. Sustainable Energy Technologies and Assessments **37**, 1006 (2020)
67. E. Tarigan, Mathematical modeling and simulation of a solar agricultural dryer with back-up biomass burner and thermal storage. Case Studies in Thermal Engineering **12**, 149–165 (2018)

Conclusions

The **methods** of the **modeling and simulation in chemical engineering** are used in the **heat energy, biotechnology** and **process systems engineering.**

Professor Christo Boyanov Boyadjiev,
Doctor of Technical Sciences,
Professor Emeritus in the
Bulgarian Academy of Sciences.

Index

A
Absolute ethanol, 21
Absorption column
    convection-diffusion model, 5
    inlet velocities, 5
    least-squares function, 7
    physical absorption, 6
    qualitative analysis, 5
Absorption columns, 5
ANSYS Fluent solution methods, 179, 182
Artificial Neural Networks (ANNs)
    advantages, 140
    architecture, 133
    artificial neuron, 130
    bias, 142, 143
    biological neuron, 130
    chemical and biochemical processes, 129
    chemical and physical laws, 129
    chemical engineering processes, 141
    data generation process, 132
    efficiency, 133
    engineering systems, 130
    genetic algorithm, 140
    growing technique, 134
    image recognition, 130
    input layer, 131
    model validation, 143, 144
    multilayer, 130, 131
    nervous-biological systems, 130
    network parameters, 133
    numerical methods, 129
    optimization, 132
    output layer, 131
    parameters, 140–142
    polymerization process, 141
    procedure, 132
    radial basis functions, 130
    recurrent connection, 132
    sensitive methods, 133
    training algorithms, 132
    training set, 132
    transfer function, 131
    types, 129, 140
    unsupervised learning, 132
    variance, 142, 143
Autocorrelation function, 144
Autothermal thermophilic aerobic digestion
    (ATAD), 136
Average-concentration model, 5, 7
    functions, 39, 42
    theoretical analysis, 40
Axial and radial velocity components
    non-uniformity, 37

B
Backpropagation-through-time algorithm, 141
Bertrand method, 27
Bioethanol, 46, 82
Biofuel production, 45
Biofuel supply chain (BSC), 49
Biomass, 46, 58, 85
Biomass-based liquid transportation fuels, 51
Biomass transportation cost, 60
Bioprocess, 148
Boltzmann's kinetic theory, 4
Boussinesq approximation, 179, 186
Buoyant forces, 179

Printed in the United States
by Baker & Taylor Publisher Services